DESIGN
ERROR
A HUMAN FACTORS APPROACH

Comment on the Cover Photo

The cover photo depicts two trains that collided with each other during a shunting episode in the Everleigh train yards, Sydney, Australia (http://www.dailytelegraph.com.au October 10, 2011). One was from Cityrail, the suburban network, and the other from the Countryrail network. Luckily no one was injured but there was damage to both trains and the accident took a significant amount of time to clear.

The movement of trains is controlled from regional control centres where controllers use highly sophisticated software to ensure that trains don't run into each other. An important control device is called 'interlocking', a procedure developed after a disastrous train crash in Armagh, Northern Ireland in 1889, where 80 people died and 260 were injured. They were on a Sunday school picnic trip and many of the victims were children.

Interlocking describes the inter-relational working of points, signals and other rail devices to ensure that once a train route is set, it cannot be reversed until the designated train has passed by.

In this photo we can see that interlocking was not employed. The accident investigation report was not released by the company so we cannot say with any confidence that this collision was caused by a design error in the control system. But we know from other rail accident reports that systems do fail when they do not include strategies to deal with every eventuality, a growing problem when automation takes over tasks previously done by humans. Computers are very good at carrying out routine activities, but have not yet been built with a human's ability to problem-solve and react quickly to an unforeseen event. We can hazard a guess that this control system had not been designed to control the whole gamut of shunting operations.

DESIGN
ERROR
A HUMAN FACTORS APPROACH

RONALD WILLIAM DAY

CRC Press
Taylor & Francis Group
Boca Raton London New York

CRC Press is an imprint of the
Taylor & Francis Group, an **informa** business

CRC Press
Taylor & Francis Group
6000 Broken Sound Parkway NW, Suite 300
Boca Raton, FL 33487-2742

© 2017 by Taylor & Francis Group, LLC
CRC Press is an imprint of Taylor & Francis Group, an Informa business

No claim to original U.S. Government works

Printed on acid-free paper
Version Date: 20160614

International Standard Book Number-13: 978-1-4987-8367-5 (Paperback)

Library of Congress Cataloging-in-Publication Data

Names: Day, Ronald William, author.
Title: Design error : a human factors approach / Ronald William Day.
Description: Boca Raton : Taylor & Francis, CRC Press, 2017.
Identifiers: LCCN 2016026982 | ISBN 9781498783675 (pbk.)
Subjects: LCSH: Design--Human factors. | System design--Quality control. | System design--Psychological aspects. | System failures (Engineering) | Manufactures--Defects--Prevention.
Classification: LCC TA166 .D39 2017 | DDC 620.8/2--dc23
LC record available at https://lccn.loc.gov/2016026982

Visit the Taylor & Francis Web site at
http://www.taylorandfrancis.com

and the CRC Press Web site at
http://www.crcpress.com

Printed and bound in the United States of America by Publishers Graphics,
LLC on sustainably sourced paper.

Contents

Foreword

A lifetime ago, I tried to replace a domestic door handle so that my knuckles did not get scraped each time I opened it. It proved impossible for the simple reason that a design decision somewhere upstream had rendered the actuating mechanism too short to give adequate clearance. This was not in any way important or life threatening, but it started a train of thought that developed into a serious interest in the way that design happens, and the – almost certainly unknown and unintended – consequences of a design process that ignores the human factor.

Human beings make mistakes. All of us. All of the time. A huge majority are of little or no consequence, except for local irritation, and pass unremarked. When the error results in a catastrophe, however, the charge is on to allocate blame. This process is marked by a search for proximal causation. Which person pressed the wrong button? Only after an exhaustive analysis has revealed that nobody at the sharp end did anything wrong, or did because of inadequate information and so forth, does the focus shift to upstream causation or contribution. Even then, when design or process inadequacies are uncovered, there is little evidence that the design process, as opposed to the outcome, is subject to scrutiny and almost no evidence at all of its being called to account.

That is what this book sets out to change. It does it, not by looking for scapegoats, but by a constructive approach to ways of doing it better. Ronald Day is uniquely qualified to do this. In addition to a penetrating (and it must be said, somewhat icono-clastic) intelligence, his background and experience allow a special understanding of a basic problem: that human actions are usefully predictable only at low resolution, and by the time the resolution gets to be useful in the design sense, they are hardly predictable at all. Murphy's law says that if anything can go wrong, it will. (If you drop your jam and bread, it will land jammy side down). Murphy's second law says that it will happen for maximum effect. (The chances of it happening are directly proportional to the cost of the carpet.) There is a little known third law that says Murphy makes mistakes. (It will all be OK sufficiently frequently to fool you.)

The practical implication of this is the need for protocols that can cope with our apparent capacity to mess up the most robust systems. This presents a challenge of majestic proportions, and it is only by a detailed scrutiny of the design processes and design thinking that any progress can be made. As automated systems driven by artificial intelligence take an increasingly important role in our critical functions, it becomes even more important to understand the human component, often at the stream source. In this book, Ronald provides a truly accessible approach to the com-plexities of safety systems design and goes on to suggest a new and important model for resolving some of the intractable issues.

It is my privilege to be associated with Ronald as a colleague and as a friend, and a great satisfaction to see this work published at a time when our rate of technologi-cal development is stretching our human capacity to cope.

Emeritus Professor Bill Green
University of Canberra

Acknowledgements

- Elaine Ouston for believing in the project and offering continued support, encouragement and the benefit of her graphic design and editing talents
- Cooperative Research Centre for Railway Innovation for funding my research
- Professor Colin Cole, Associate Professor Yvonne Toft, Dr Ryan Kift and those others at Central Queensland University who offered inspiration and suggestions

Introduction

Good design is good business.

Thomas J. Watson Jr

If I asked you to describe what a designer does, many of you would respond with a visual example. We tend to think of designers as working in *haute couture* or in graphic design, building houses or bridges or cars. But, design goes so much further. People like Steve Jobs design computers and other electronic equipment. Engineers design everything from railway tracks and trains to planes, robots and new mousetraps.

Information and communication technology designers create new computer programs to do everything from making your microwave work to telling the different components of your car engine how to operate together to guiding a spacecraft as it blasts off into orbit around the earth, or flies to the moon or Mars. Computer programs are everywhere, in banking and finance, signalling and control systems, film animation and global positioning guidance systems. Computers now drive trains, fly planes and guide trucks and coal cutters in mines. They have controlled the mechanical and electronic operation of cars for some years, but now they are beginning to take over the driver's seat, a move that is accelerating fast.

But designers also work in factories and businesses, and almost every other occupational situation, designing the work processes and procedures that guide the people who get the job done. Hopefully, these processes and procedures will also produce safe and productive workplaces, but we know that does not always happen.

A significant number of incidents, accidents and fatalities in homes, workplaces, transport and other places where people gather can be blamed on design error. People fall from poorly designed ladders and unsafe high-level gangplanks and gantries. Decks attached to houses collapse, trains crash, planes fall from the sky, electrical equipment shorts and starts fires and information and communication technology systems fail. In the first instance, human error is normally blamed, but often the primary cause of an accident can be traced back to design error.

In business terms, design errors cost money. Where design error is identified, the responsible company may have to pay millions in compensation. Products may have to be recalled and retrofitted at a serious cost of time and money. Processes and systems have to be modified and time wasted.

But those time and budget blowouts can be avoided by the application of a few simple rules during the design process of the product, process or system. This book reports research into the design process, particularly of information and communication technology systems, engineering products and business processes and procedures.

The research looked closely into the design processes used by these designers. Crucial steps and potential safety hazards are identified, and recommendations made to mitigate design error. Those recommendations are being fashioned into new problem-solving strategies and an analytical tool which will assist designers, managers and business process and occupational health and safety personnel in identifying both safe and potentially hazardous behaviours in the various stages of the design process. These strategies and tools can be used for systems auditing and training and to guide accident investigators.

Many design errors occur because of the poor choice of a systems development life cycle (SDLC) model to guide the design process. An organic view of the design process is described. It brings a new vision to the design process as a complete entity. A new continuous Agile SDLC is proposed to help speed up the design process and adapt to the increased complexities and rapidly changing work and business environments evident around us, while at the same time producing safer and risk-free technologies, devices and work processes.

This book provides a research-based investigation into the design process, pointing out how and where errors creep in, and offering practical advice on strategies to remove potential hazards from design projects in the future.

The rapidly advancing complexity in systems, devices and processes brings new challenges to bedevil designers. Errors in designs for the brave new world that we find ourselves in threaten to cause chaos. Accidents, incidents and disasters will have greater impact on larger numbers of people than ever before. The Chernobyl and Three-Mile Island disasters will look puny in comparison with the potential hazards of the future if we do not devise new ways to error-proof our designs.

Without the ability to identify and mitigate design error in complex systems, large companies, governments and nations will collapse. This book provides you with a detailed description of the ways design processes have been carried out in the past and suggests a range of strategies that need to be adopted to mitigate design error in future complex systems. Real-life examples are used to demonstrate the points being made. Many of the concerns raised in the book have come from a worldwide study conducted with designers, managers and end users in 2011.

CHARTING THE DIRECTION OF THE BOOK

CHAPTER 1: DESIGN ERROR ... CAN MODELS HELP?

This chapter discusses some of the more common models of accident causation that have been used to describe the development of design error in systems. Heinrich's domino theory is looked at first, followed by Reason's Swiss cheese model, Hollnagel's systemic view and a new random clusters model developed by the author of this book.

Each of these models is used to explain the design issues that led to a significant accident. Those accidents are described in detail and the reasons for each accident explained. They include the Queensland Health payroll fiasco, the Zanthus rail crash, the Air New Zealand Mt Erebus disaster, the Australian government's Home Insulation Program deaths and the Melbourne West Gate Bridge collapse.

CHAPTER 2: COMMON CAUSES OF DESIGN ERROR

There are many causes of design error, as this chapter will outline. To demonstrate one of the big ones, designer lack of user situational knowledge, the author relates the story of his purchase of a lawn mower.

The term 'client' is introduced to describe the managers of organisations who engage design teams, and reference is made to their lack of ability to describe in close detail the operational requirements of end users. This introduces a second major cause of design error, too much reliance on client documentation without validating premises with end users.

Other causes of error described in this chapter include user interface or usability issues, and the important discussion on design constraints.

Chapter 3: Who Is the Designer?

To understand the context of the design process, we need to study the designer. This person or team sits in the middle guiding design process activities.

The first discussion revolves around how a designer creates a new design. Some designers say they use intuition, while others hold that they employ researched design strategies, brainstorming and 'divide and conquer', where large complex problems are broken down into smaller solvable units. A designer profile is offered that is a list of the commonalities gathered from the research. Although just a generalised list from the 2011 study, it does throw some light on the people who do much of the design work in our societies, and goes a long way to helping us understand why end users are so often left out of the design process when they have so much localised knowledge and experience to offer.

Chapter 4: What Is the Design Process?

This chapter looks in detail at each of the steps that must take place during the development of a design. Each of the seven steps is dissected minutely to make it abundantly clear what needs to happen at each phase of the design process. Those steps are

- Design concept formation
- Writing specifications
- Building the design
- Testing
- Implementation
- Training
- Maintenance

Chapter 5: Systems Development Life Cycle Models

The seven steps outlined in Chapter 4 form part of every design process, but they vary in the way they are carried out and the order of precedence and importance depending on the SDLC model that is chosen to guide the design team.

SDLCs were first used in the information and communication technology industry more than 50 years ago, but they have been adopted and adapted by designers in many other fields over the intervening years. This chapter describes the major SDLC models and comments on their strengths and weaknesses. They have been divided into traditional and adaptive models, and the differences between those two groups are explained.

Chapter 6: Choosing an SDLC ... Which Cap to Wear?

The SDLC models described in Chapter 5 are only a few of the available choices, but they are the most popular with designers around the world. There are many localised modifications of these.

The choice of an SDLC is not an easy one, and depends very much on the design task. In broad terms, the design team needs to choose one of three options, depending on the design brief. The author has chosen to describe each in terms of wearing a different coloured cap.

The first choice is the RED CAP option. RED CAP SDLCs are traditional waterfall-style models that are highly suited to those big projects like designing a high-rise building, constructing a large bridge or sending a rocket into space, but they are slow and cumbersome, allowing no input from stakeholders and end users, nor flexibility to modify an earlier part of the design. The author outlines a mandatory modification that this model requires to provide some flexibility and make this model more responsive to stakeholder and end-user input. Once this beast trundles to the starting line, it becomes irreversible and unstoppable until the project is complete.

BLUE CAP projects sit between traditional and adaptive solutions. They are prototypes where an idea is built and tested before becoming a YELLOW CAP project.

YELLOW CAP methodologies are ideally suited to most design projects. They are adaptive in nature, which implies that stakeholders and end users have membership of the design team. These projects, when carried out according to adaptive rules, are normally completed more quickly, cost less and are less likely to contain errors.

Chapter 7: Where Do Design Errors Occur in the Design Process?

The most risk-prone design process stage is the very first step, the formulation of a design concept. The need for a design concept arises because of a perceived problem or issue that needs to be amended, or a new tool or process created.

A design concept needs to take into account every factor, every variable associated with successfully solving the problem. When a design concept is flawed, the rest of the design process is doomed to failure, so the challenge is to arrive at a concept that will lead to a successful resolution.

Sadly, many design concepts are flawed and the resulting design either fails to solve all aspects of the original problem or, worse still, allows design errors to occur during the design process that can lead to accidents and deaths.

Beyond the concept formation stage, errors can creep into a design at every step. Each design process step is investigated and potential error sources identified.

Chapter 8: Human Factors Issues

This chapter introduces the author's disconnect model, which shows how little attention is given by management and designers to the end users. It also indicates the greater level of communication between designers and clients and shows how neither group communicates fully with the end users, those people who know the climate, deal with the day-to-day stresses and strains and can give precise operational detail to the designers. It explains where mistakes can be made at a systems or organisational level.

In this chapter, we also hear from end users. Some of their responses to research questions are examined. From these responses, we are able to draw up a list of some of the user interface and usability issues that should be taken into consideration by design teams.

CHAPTER 9: AUTOMATION ... PERSISTENCE OF A MYTH

Dekker (2011) describes the drift into failure when mechanical or electronic equipment begins to exhibit stress and fatigue that is not identified by normal maintenance procedures. When a part in a plane, train or control system fails, it can come as a complete surprise, and often the failure is blamed on human operator error when really it should be regarded as design error or maintenance error. The end user, the operator, is often the first to be blamed instead of the last. This person is a convenient scapegoat.

Because of the persistence of this myth, huge strides have been taken in removing humans from machine operation to reduce the likelihood of human error. What is not recognised is the greater potential for error from the resulting highly complex automated designs.

For years, we have had an automatic pilot that can be turned on to take some of the drudgery out of flying, but now we have not only drones, but also passenger aircraft that do not have a pilot at all. The 1988 Airbus A320 crash at the Habsheim air show in France offers a classic example of what can go wrong when the pilot is designed out of the cockpit.

Currently (2015) there are 206 automated rail systems operating around the world, some with 'token' drivers and some without. The disastrous automated high-speed rail crash near Wenzhou, China, killed 40 and injured 200. A Spanish automated high-speed rail disaster crash killed 80 and injured 130. Automated trucks and underground coal cutters are now appearing in large mining operations. Now, we are seeing the world's first driverless cars (Yarrow 2014).

Design difficulties with automated systems are described with the aid of the author's levels of complexity diagram.

The chapter ends with a detailed examination of the recent Internet of Things explosion and the risks this technology brings with it. Examples are given of large companies that have been targeted and had credit card details stolen from their point-of-sale systems, hackers often entering these companies through their air conditioning or other Internet of Things systems and then gaining access to their point-of-sale systems.

CHAPTER 10: HOW ARTIFICIAL IS ARTIFICIAL INTELLIGENCE (AI)?

Artificial intelligence is a joint traveller with other automated technologies, and will eventually outstrip and control them all. In the 1970s, the first attempts to develop artificial intelligence systems were thwarted by lack of memory. Asking a computer to act like a thinking being requires an enormous amount of processing power, and those early computers just did not have the 'grunt'. Improvements in computer speed and storage space allowed researchers to experiment with ways in which a computer could be made to 'think like a human'.

The relentless march towards producing a machine that can 'think' now threatens life as we know it. Steven Hawking believes that highly automated 'thinking' machines might eventually replace the human race. Some of the potential problems centre around the lack of emotion and 'humanity' when these machines make decisions that will affect us, and the increasing complexity of both their design and their maintenance that may eventually mean humans can no longer understand how they work.

CHAPTER 11: THE SOLUTION IS ...?

This chapter begins by summarising the many ways that a design process can go wrong, design stage by design stage. Human factors considerations form a large part of this summary.

The chapter then lists and explains a series of rules to guide designers as they go about their work. The rules will vary depending on whether we are considering RED CAP, BLUE CAP or YELLOW CAP projects.

The final section is the description of an analytical tool being developed for designers, managers, auditing staff, accident investigators and anyone else who has a need to investigate the design process. This tool, the Design Error Avoidance Model, will not only help with the assessment of design processes, but also provide a basis for training staff in error reduction strategies, and provide accident investigators with a tool to help them determine the causes of incidents and recommend strategies to avoid design error problems in the future.

Author

Ronald Day is a lifelong author, teacher, theorist and researcher. He was one of the first human factors professionals, encouraging designers to consider the ways users interact with the systems, devices or procedures they develop.

Ronald has worked in universities, government organisations, information technology businesses and power production and distribution companies and on mine sites. Much of his work has been with information and communication technology and business departments within those organisations. Currently, he is an occasional lecturer at Central Queensland University, Australia, and a lead consultant for Safe Design Solutions. In every role, he has looked for the design errors, the flaws that could cause incidents, accidents and perhaps even deaths. He has worked on two fronts, one to remove faults before they become issues, and the other to ensure that those systems have been built with end users in mind. More information can be found on Ronald's website – www.safedesignsolutions.com.

1 Design Error ... Can Models Help?

Design must reflect the practical and aesthetic in business but above all ... good design must primarily serve people.

Thomas J. Watson

How many times have you bought a new electrical appliance, plugged it in and nothing happened, or there was a blue flash before it died? You find an unexplained charge on your bank statement and you are told that their new computer system made a mistake. You are assembling a flat-pack piece of furniture when you discover that the instructions miss a vital step in the assembly routine. All of these indicate an error somewhere in the design process.

In simple terms, a design error is an aspect of a design that, when put into operation, does not work as it should. This flawed aspect can cause annoyance for users or, more seriously, can cause incidents, accidents or death on the part of users or the general public.

Errors in the design process have been described through accident causation models. One of the earliest models is H. W. Heinrich's domino theory, where the tiles fall down one after the other. This sequential model implies that any accident has a single cause, whereas in fact most accidents have more than one contributing factor. Nevertheless, it provides a useful model for studying the dynamics of the Queensland Health payroll fiasco that we will look at in more detail later. Suffice it to say, at this stage the first domino, the formation of the design concept, fell over because the designers did not talk to the health payroll people to find out the true size and complexity of the problem. The first domino fell over, knocking down the second. The second domino knocked down the third, and so on. Because the concept was flawed, every stage beyond that was also flawed, and when installed the system did not work. Eighty-five thousand staff did not get paid that fortnight, many of them for much longer than that. The falling domino theory and Erik Hollnagel's sequential model too (Hollnagel 2004), describe the linear and deterministic progression of the flaw through the whole design process.

In the 1980s, epidemiological accident causation models began to appear. These models regarded 'events leading to accidents as analogous to the spreading of a disease' (Qureshi 2007). James Reason's organisational model (Reason 1995, 2008), better known as the Swiss cheese model, describes those accidents where a number of factors line up, some due to active failures and some due to latent conditions. This combination of error factors, still linear but more organisational, eventually causes an accident. Besnard and Baxter (2003) want to add a technical dimension to the Swiss cheese model. They believe that considering technical aspects along with the organisational story provides a fuller explanation of the error flow that can course through a design. Qureshi (2007) suggests the holes in the Swiss cheese are constantly moving. He holds that Reason's model reflects a static view, whereas the real situation is more dynamic.

For this model, I have chosen an engineering example, the Zanthus rail crash which happened in Western Australia near the South Australian border. I explain the detail of this accident a little later in this chapter. In brief, an engineering firm was commissioned to design and build a solar-operated switching device to allow points on remote sections of railway track to be operated from a train. One train was resting in a bypass loop, while another train passed, travelling in the opposite direction. The button was pushed at the wrong time, and the travelling train entered the bypass loop and crashed into the resting train. The technology worked, but because the designers did not do their research, a number of hazards lined up. The designers did not speak with train drivers to understand the operational setting. They did not build in any interlocking, a standard practice for well over a hundred years of rail operation; they did not build in any time delays; their controls were not labelled; and no written instructions were provided. These factors lined up like holes in a number of slices of Swiss cheese, making an accident inevitable.

Hollnagel (2004) has suggested we need to take a systemic view when considering more complex technological designs. In this view, we need to consider a wider causal base, taking into account human, technical and environmental factors. This model of accident causation can be seen operating in the Mt Erebus crash of an Air New Zealand flight taking sightseers to Antarctica. Office-based technical staff followed a management decision and changed the route by re-programming the automatic pilot in the plane. Unfortunately, they omitted to communicate the change to the flight crew and 257 people died when the plane crashed into the side of the mountain.

But there are some design-induced errors that cannot be explained by the falling domino theory, the Swiss cheese model, or the systemic view. We need a new model that explains those design error situations where a number of dissimilar dysfunctional strands run through the design process and gather into clusters at the wrong time. To examine this new type of accident causation model, we will explore two disasters: the first being an Australian government program that went terribly wrong and led to several deaths and the second a huge bridge that collapsed during construction and killed many workers. We will call this new model the *random clusters* model (Figure 1.1).

Figure 1.1 describes the way hazard elements can cluster with other hazard elements in a random manner to cause a succession of crisis points within a complex project. Each of these crisis points can contribute to the failure of a project, one building on another, but not in a linear sense like the previous models we looked at. This model operates in a three-dimensional manner, with hazards linking up with other hazards with cross-over effects in time and space.

All the previously discussed models work well within a clearly organised and orchestrated environment, but they fail to describe accidents where political power plays and public service obduration, a seemingly endless pot of money, a large number of untrained and inexperienced contractors and workers, a degree of rorting, a lack of health and safety observance and a lack of supervision come together. The random clusters model is the only model I can use to describe the calamitous and

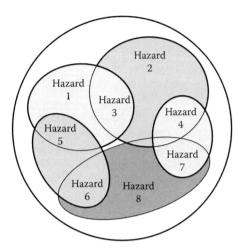

FIGURE 1.1 Random clusters accident causation model. Model devised by the author to represent the manner in which operational elements of a complex project can be grouped in dysfunctional clusters.

chaotic Australian government's Home Insulation Program, a scheme designed to fit roofing insulation into upwards of 2.2 million homes to improve home heating and cooling and reduce power consumption. The Royal Commission Report (Hangar 2014) admits the scheme was well intentioned. The Rudd government looked for a way to lift Australia out of the global financial crisis (GFC), but sadly the scheme wasted millions of public dollars, was subjected to rorting, caused almost 200 house fires, cost the lives of four young men and severely injured a female insulation fitter.

Another government project that failed with disastrous consequences, this time in the Australian state of Victoria, also can only by described in terms of the random clusters accident causation model. A tunnel was proposed by a panel of government advisers to improve the traffic flow from Melbourne's inner city centre to the western suburbs. The government decided it wanted a bridge, this one to be higher and longer than the famous Sydney Harbour Bridge. There has always been rivalry between Melbourne and Sydney, and the Victorian government saw this as an opportunity to score some points. The government departments were at loggerheads, the numerous contracting companies could not work together, an inappropriate bridge model was chosen, safety considerations were ignored and part of the structure collapsed, killing 35 workers.

DOMINOES ARE FALLING: QUEENSLAND HEALTH PAYROLL FIASCO

The Queensland government commissioned a new payroll system for the state's health professional body. In this state, all public health services are delivered by a single government department. At that time, the 85,000 staff formed a diverse group including people at every level, from surgeons and anaesthetists to doctors and nurses, psychologists and physiotherapists, ambulance personnel and other para-professionals, accountants and administration staff, hospital orderlies and cleaners.

Many of those people worked 24/7 shifts. The KPMG (2012) report states that these workers were covered by 12 industrial awards and impacted by 6 different industrial agreements. An additional 200 separate allowances resulted in more than 24,000 different pay combinations.

The consortium that gained the contract to build and install the payroll system originally quoted $A6.1 million. They used as a model a system built for 1900 workers on two salary scales, a seriously inadequate model to use. Because of the complexities, a series of delays caused the new system to be delivered two years late. Because of the delays and the government's urgency to have this new system operating as quickly as possible, no full system testing was done. The old system was turned off, the new one was turned on and it failed to work. No one was paid that fortnight. A large team of people was engaged to do manual pays so that people could be paid the next fortnight. They could not cope, so the pay cycle was pushed out to once every three weeks. Some received no pay for longer than that. Some health workers were still not being paid correctly more than two years after the new payroll system was introduced.

An early Australian Associated Press (AAP) independent report stated it would take 18 months and $209 million to fix. A later report stated that the government would spend $1.25 billion by the time the payroll system worked correctly, not including what would be spent on the physical upgrading of the system.

The official KPMG auditors' report identified that the flawed design concept was caused by no design team members communicating with the end users, the payroll staff at Queensland Health, to determine the requirements of the system. The report also slated the lack of thorough testing and the lack of adequate training of payroll staff in the new system.

The new health payroll system was installed without a full end-to-end test and turned on. The old system was turned off with no comparison runs done. Normally with an installation like this, the new system would run in dummy mode alongside the old system and the results would be compared at the end of the run. This might be carried out four, five or more times until it was proved the new system would operate correctly. At this time, the old system would be turned off and the new system moved into 'go-live' mode.

So in this case, because the design concept was flawed, that problem was carried through every following step in the design process. In addition to the problems caused by testing against flawed specifications, no whole system testing was done and insufficient training was given to the payroll staff, additional issues that compounded the disaster. This is a classic example of Heinrich's domino model in operation.

SWISS CHEESE ANYONE? ZANTHUS BYPASS ACCIDENT

On August 18, 1999, a freight train heading for Sydney waited at the Zanthus bypass in Western Australia near the South Australian border. The section of track crossing the Nullarbor Desert is a single railway line with bypass loops every 30–60 kilometres. This section of track is too far from a control centre for remote control of the points that need to be moved to allow entry into the bypass. To enter a bypass in the past, a train had to stop to allow an engineer to demount and shift the points manually. To speed things up, solar-powered wireless devices were designed and installed so that points could be shifted from the train.

The second engineer on the freight train was waiting for the Indian Pacific passenger train to pass on its way from Sydney to Perth. The control box was unlocked and open, and his finger was poised on the button. He pressed the button too soon and the Indian Pacific was diverted into the bypass loop, where it crashed headlong into the freight train.

Forty-five people were injured. Twenty-one of them were flown by Flying Doctor to the Kalgoorlie Hospital. The crash cost the railway companies upwards of $A10 million.

The accident investigation team found the design of the wireless device was defective through having no interlocking to prevent the points being moved before the passenger train had passed. They also found the interface unacceptable because there was no labelling to show which button did what. They noted the human error where the engineer pushed the button too soon, and recommended a time delay be added to the design.

The accident report showed three major reasons for the accident. The first cause was the engineer pushing the button at the wrong time. This can be seen clearly as a *human error*. But the bigger question is, why did the new system allow him to press the button too soon?

Currie (1971) reports that in 1889, a horrific rail tragedy in Armagh, Northern Ireland, resulted in 80 people on a Sunday school excursion, many of them children, being killed and 260 injured. Since that time, railway companies have built into their systems what is termed 'interlocking'. This term describes the interrelational working of points, signals and other rail devices that ensure that once a train route is set, it cannot be reversed until the designated train has passed by.

It is clear that the designers of the wirelessly controlled points changing system had no knowledge of normal rail protocol when a choice of routes is available. It would not have taken much research to discover what has been normal rail practice for more than 120 years. This type of design error is caused by a *lack of industry operational research* relevant to the device being designed.

It is also clear the designers had no vision of the human factors issues surrounding workers and their operating environment, nor did they have any knowledge of user interface design. The switching box had two buttons with no labelling to indicate which switch did what. There were no instructions or directions to guide users of this system. How could any designer imagine that a user would instinctively know what the buttons in this box would do? How could they assume it was not their responsibility to provide operational instructions? This describes another type of design error that we can attribute to a *lack of user interface or usability awareness*.

We know the engineer pushed the button too early. We also know there was nothing built into the system to prevent him from doing just that. The fact the designers did not consider this could happen leads us to declare another type of design error. This one we might call a *lack of user situational knowledge*. This problem could have been solved easily by building in a time delay to disallow a second button being pressed before the other train had passed.

There is another scenario we should investigate. Not all blame can be levelled at the designers. They were employed by rail company executives. We will call these people 'clients'. In my experience, clients are often as divorced from the real situation as the designers. They are the people who engaged the design company to design the new wirelessly controlled apparatus. The clients wrote the requirements to guide the designers when creating their design concept. Their knowledge of history should have been greater, so they should have specified that the design needed to include some form of interlocking. But I doubt we can expect their knowledge of human factors issues and situational awareness to be any better than that of the designers. So we can assume the original contract documents were flawed to the extent that they did not mention interlocking or user interface design.

Although probably not mentioned in the original specifications, we cannot let the designers off the hook too gently. If they had taken a ride on one of the freight trains and talked to the engineers, they should have had a much clearer vision of how their design would be used in the operational setting. If they had tested the new system thoroughly in a typical situation before releasing it to all trains, there would have been a much smaller chance of operational errors remaining in the design. We can guess the designers made another design error, one that can be described as *placing too much reliance on client documentation without validating the problem with the end users.*

This accident fits Reason's Swiss cheese model very well. A number of factors lined up like a series of holes in slices of cheese. When that happened, the accident was inevitable.

SYSTEMS FLYING HIGH: MT EREBUS DISASTER

On 28 November 1979, an Air New Zealand flight set off to take a party of 237 sightseers to Antarctica. The normal flight took the passengers up the 60-kilometre-wide McMurdo Sound, where the plane dropped as low as from 1500 feet down to 600 feet so all could see the ice, the buildings and perhaps teams of dogs pulling sleds. They might even see the huts of the early explorers, Scott and Shackleton.

Brown (2008) tells us the flight path was computer generated and loaded into the plane's automatic pilot. Without telling the crew, the planners decided on a change to the computerised flight path. Instead of going over McMurdo Sound, the plane would fly 43 kilometres further to the east directly over Antarctica's highest peak, the 3794-metre-high Mt Erebus.

That day, dense white cloud obscured any view of the land from the normal cruising height of 35,000 feet, so the plane descended to 16,000 feet. Cloud was now down to 2000 feet above McMurdo Sound and light snow was falling. As the plane descended further, it encountered what is called 'whiteout', which occurs when cloud layers diffuse the direct rays of the sun, making the terrain look like sea ice. In those conditions, it is impossible to define the land surface and judge distances.

The plane disintegrated as it hit the mountain, and all on board were killed instantly. Nobody had bothered to tell the crew of the change to the route.

When Air New Zealand top management discovered the route had been changed by the flight navigation section acting on flawed information from their superiors, a huge cover-up ensued. It was claimed that airline staff began to shred documents.

The chief inspector of airline crashes laid the blame on the flight crew, particularly the first pilot, Collins. The New Zealand government announced a royal commission chaired by Judge Mahon. He viewed around 900 photographs recovered from the accident site by investigators, including one taken at the time of collision showing clear air. Mahon went to Antarctica and flew around Mt Erebus, by chance experiencing the same kind of whiteout that had confused the flight crew. His report exonerated the flight crew and directed the blame squarely at Air New Zealand management.

This is a classic case of designers implementing a change and not bothering to communicate the change to the end users, in this case the aircrew. That lack of communication cost the lives of 257 people. Again, we can see a linear progression of faults. Management personnel changed the flight but did not inform the crew who had taken this flight many times and expected the automatic pilot to act as it had done on their previous flights. The captain saw the mountain approaching too late. He tried to pull the plane up to fly over, but it was too late. Too many people died because of a management decision that was not communicated to the crew. Hollnagel's systems model of accident causation fits this accident very well.

RANDOM CLUSTERS MODEL: GOVERNMENT HOME INSULATION PROGRAM

In 2009, the Australian Rudd government cast around for a program that would employ a lot of people and help stave off the potential recession that could occur because of the global financial crisis (GFC). Within the $A42 billion stimulus package was a $4 billion allocation to employ contractors to install ceiling insulation in the roof space in upwards of 2.2 million houses. It was held that this would decrease heating and cooling costs and lower the consumption of power, at the same time providing a significant number of jobs.

The consequent Royal Commission Report (Hangar 2014) informs us that before the scheme began, there were approximately 200 registered ceiling insulation fitters. Once the program began, the number grew to around 8,359 businesses with a total workforce of around 12,000. Many of these businesses had little or no experience in the installation of home insulation. Young, inexperienced people were hired to climb around in ceilings with no training and no explanation of the safety risks.

Those ceiling spaces were crammed with power cables lying haphazardly across ceiling supports and gyprock sheeting. In many buildings, older copper wire was surrounded by corroded insulation. Many of the contractors were using aluminium insulating material that was placed on top of the wiring and stapled, often with steel staples. Few contractors bothered to turn off the power to the house before entering the ceiling space. Two young men were electrocuted while using staple guns to fix the aluminium sheeting, penetrating live cables as they did so. Another worker was pushing fibre insulation material into a tight space with a metal rod and was electrocuted when his rod touched a live wire. A fourth young man died from heat stress when working in a ceiling space on a hot day with no drinking water. A young woman who attempted to help one of the electrocuted men was severely injured.

In addition, 197 house fires were reported as being caused by the Home Insulation Program.

The Royal Commission identified seven significant design failings:

1. Conflict of aims: The first was a conflict between the two aims of the program. On the one hand, the government intended to insulate 2.2 million homes. On the other hand, the program was designed to stimulate the economy. The first of these aims required careful planning and execution over a long period of time. The second aim required speed. Consequently, the program was begun in haste and essential precautions were not taken.

2. Locus of control: The second thread relates to the allocation of the program to the Department of the Environment, Water, Heritage and the Arts (DEWHA), a body the Royal Commissioner believed 'was ill-equipped to deal with a program of its size and complexity'. That decision was exacerbated by changing from a regionally managed model, as suggested by DEWHA, to a direct-delivery mode, which it was said would speed up delivery.

3. Risk management: The third serious issue was the failure until late in the program to identify and manage the risks to installers.

4. Dangerous product: The fourth issue was permitting a dangerous product to be used by contractors. Suppliers and building regulators indicate clearly that foil laminate–type products should never be laid on top of ceiling joists and plasterboard, nor over down lights, because of the danger of coming into contact with electrical wiring. Two workers died and one was injured when steel staples were used to secure the aluminium sheeting to timber joists. They penetrated electrical cables and the current was transferred to the sheeting material.

5. Lack of training: The fifth design failing was relaxing training and competency requirements so that they could be replaced by 'supervision' in the short term. The supervision aspect was never specified and so fell by the wayside in the rush to get a large number of unqualified contractors into the program.

6. Lack of audit: The report lists as its sixth failing the lack of a 'robust audit and compliance regime' before beginning Phase 2 of the program. No serious auditing was done until after the first death.

7. Poor communication: The last item on the list was the reliance upon the states and territories 'to regulate, monitor, police and enforce' occupational health and safety management by contractors. The Royal Commission reported that the Australian government never made its expectations clear to the states and territories; neither did it check they had the resources to manage the program in their jurisdictions.

These streams ran separately, but at times converged in groupings of two or more strands to heighten the degree of risk. The deaths of two of those four young men and the serious injuries to a young woman occurred when a dangerous product was used without training, and when no risk management strategies were offered by the contractors. In this cluster were overtones from the aims conflict, locus of control issues and lack of an audit.

One young man died because of the clustering of the lack of risk management and training and the use of a dangerous tool. The risk was compounded by the lack of an early audit of the program by responsible experts, the clash of aims and the lack of communication between the federal and state governments.

The fourth death occurred because of the clustering of environmental conditions and the lack of risk management and training from the contractor.

CHAOS REIGNS: MELBOURNE'S WEST GATE BRIDGE COLLAPSE

Another disaster that can only be classified in accident causation terms as following the random clusters model is the collapse of Melbourne's West Gate Bridge that caused the deaths of 35 men (Brown 2008).

The West Gate Bridge in Melbourne's western suburbs was planned to open up the west to business development and housing. The project began in turmoil when the government over-ruled its own committee and opted for a bridge instead of a tunnel, which would have been a cheaper and quicker solution to the required movement of traffic. One of the government's secret aims was to provide Melbourne with a world-class structure higher and longer that the Sydney Harbour Bridge. So the conflict within government departments set the tone for the whole project.

The 'few' houses that needed to be demolished to make way for the bridge turned into 150, to the annoyance of the families that needed to move.

Sixty companies were involved in one or more aspects of the project, and serious problems developed between them. Four of those companies were steel contractors, and issues between them turned into industrial disputes and stoppages. Union demarcation disputes stopped the project on many occasions.

A serious design error factor was the adoption of a box-girder construction where a number of sections were assembled on the ground and then lifted up to the concrete pylons, one-half of the span at a time. Each of these spans was more than 100 metres in length and weighed around 1400 tonnes. This construction method produced sagging and buckling, and these problems created enormous issues when the second side of each span was lifted. Bolting these two sections together did not go smoothly, and several remedies were attempted to reduce buckling so the two sides could be matched. Some of the more extreme solutions involved unsafe handling procedures.

Contractors lied about buckling and other problems, but eventually the Bridge Authority felt it could not trust the assurances it was receiving from the contractors and called in a London firm of engineers to check the bridge structure. The audit report listed 14 'inadequacies' in the steel section, but the report was not released at the time. Had those findings been made known, there may have been changes in construction methods and the collapse may not have occurred.

Problems began when the contractors attempted to lift the first half span between piers 10 and 11 on the west side. When the second half span was lifted, it was found that the camber on the northern half was 10 centimetres higher than that on the southern half. The builders lifted ten 80-tonne blocks of concrete to attempt to level the higher side. As the last of these blocks was being put into place, a buckle appeared near the centre of the span. It was a big one, more than 3 metres long. The two spans were still on their temporary bearings, a precarious situation. One of the bosses ordered the removal of some of the bolts holding the two spans together. After 30 bolts had been removed, the buckling flattened out and then a serious buckling spread through the whole length of the two spans. Workers tried to get bolts back into place as quickly as they could, but they could not halt the movement. Flakes of rust peeling off the steel caused an eerie pinging noise. Then the bolts began to turn blue. The whole 100-metre span fell 50 metres to the ground, taking the workers with it. Thirty-five died. It is reported that others died from stress following the disaster.

Warning bells had begun to ring when box-girder bridges in other countries began to fail. The Fourth Danube Bridge in Vienna sagged up to a metre in places. The Milford Haven Bridge in Wales collapsed during construction and killed four men. After the collapse of the West Gate Bridge, 61 bridges around the world were closed pending inspection and strengthening, and another 41 had reduced traffic flows while checks were made.

This calamity was the result of the random clustering of a number of different risk factors. In this mix, we have government departments at cross purposes, too many contract companies not working well together, serious industrial issues with demarcation disputes and strikes, a risky structural method, contractors hiding problems they were experiencing from officials, potentially dangerous solutions to unforeseen problems and a lack of safe processes and procedures.

***

Now that we have looked at some of the accident causation models, it is time to investigate the common causes of design error. There are many, as Chapter 2 outlines. To demonstrate one of the big ones, *designer lack of user situational knowledge*, I relate the story of my purchase of a lawn mower. How can the purchase of a lawn mower help us understand this important principle? Read on and you will soon find out.

2 Common Causes of Design Error

Does human error cause accidents? Yes, but we need to know what led to the error: in the majority of instances it is inappropriate design of equipment or procedures.

Donald Norman

I recently bought a new lawn mower. *So what has that got to do with design error?* I can hear you ask. I looked at the assortment of machines on display and finally settled on two. Choosing between them was difficult. One was a little cheaper; the other had a slightly more powerful engine. Then I imagined myself pushing the mower around the lawn and thought of the ways I interact with it. But of course, the thing I do most often when mowing is empty the grass catcher. I bent down and unhooked the catcher on the first mower. It lifted away easily. I looked at the way it fitted to the machine and noticed the two simple hooks that were designed to sit on a bar above the rear opening. It was immediately clear to me exactly how and where to fit it. I replaced it in a second – easy! The human factors issues surrounding person plus machine plus operational environment had been clearly thought through and designed into the device. Full marks to that designer!

Then I tried the second machine. I lifted off the catcher and looked at its attachment mechanism. What a complicated assortment of bars and holes and bent metal it presented. I tried to fit it back onto the machine. There was no clear indication of where it should attach – no simple one-to-one alignment of hook and bar like the first machine. I pushed and shoved and wiggled. It caught on one side and dropped lopsidedly on the other. I took it away and tried again. It took me several minutes to attach it firmly in place. *Whoever designed this did not put himself or herself in the place of the user*, I thought.

The designer solved the problem of designing a grass catcher that could be attached to the mower. To that extent, the design met the specifications. But the design was not user-friendly. It was clumsy and lacked transparency and usability. It failed to address the user's conceptual framework – the way users think about and approach the task for which this device was designed. I have a large lawn area and I need to empty the catcher maybe 9 or 10 times on each occasion I mow. I want a catcher I can detach, empty and re-attach in the shortest time possible. I certainly do not want a catcher that takes me three or four minutes each time to puzzle out how to re-attach it. If the designer had taken a prototype of this machine around the district and invited neighbours to try it out, I feel sure the difficulty would have been identified quickly.

VALUE OF USER SITUATIONAL KNOWLEDGE

Often designers begin by looking at previous similar projects to determine whether one of these can be used as a model for the new design. When a previous design is chosen as a model, all too often it is not checked against the operational needs of the new problem. Lack of *operational research* is often a contributor to design error. In my lawn mower example, the designers would have used their time well if they had allowed some neighbours to try out the new design before it went into production. In the Zanthus example, if the designers and rail managers too had travelled on a freight train on that line and talked with the engineers, they would have understood the operational needs much more clearly.

CLIENTS ARE PART OF THE PROBLEM

Design normally begins with a problem to be solved. The problem is often stated in a contract drawn up by clients. As we shall see later, the clients, that is, the managers in large organisations who commission the design, often do not verify the operational setting and requirements with their end users. In the study that led to this book, less than 25% of designers talked with end users to verify how the design would be used in their work. In many cases, their view was as distant from the workface as the clients.

TOO MUCH RELIANCE ON CLIENT DOCUMENTATION

As we have discussed earlier, the people we call 'clients' in this discussion are those who engage the designers and draw up the contracts or specifications for new or reworked technologies, systems and processes. Too often, particularly in large organisations, these people are far removed from the end users, the people who need to implement the new design in their work. These new designs can be new information and communication technology systems, technical or engineered devices, or business work processes and procedures.

Clients may have a very good general knowledge of the business, but they are often not fully aware of the nitty-gritty of operations, and often do not consult with the end users to determine the finer detail of the tasks done by these people. Many of them are not aware of the human factors issues, including the problems that can occur when the user interface is poorly designed.

The contracts and specifications these people draw up may not reflect the operational requirements exactly. When designers work from these flawed documents, the final design will also reflect these flaws and carry potential safety issues, particularly for designs that will be used by people in safety-critical situations.

We saw this earlier with the Zanthus railway disaster, where the specifications from the clients carried no information on interlocking or user interface design requirements.

USER INTERFACE OR USABILITY ISSUES

Many designers are not knowledgeable about user interface design. We have all experienced situations where the operation of an appliance is confusing and not well described in the handout or manual. Older readers will remember the problems many of us had with early video recorders.

It is more common than it should be when assembling flat-pack furniture to find unexplained or missing steps in the assembly instructions. Computer systems and Web pages can annoy us because of poorly laid out screen information or confusing menu choices.

Chapanis (1996) described user interface error as the major cause of the Three-Mile Island nuclear disaster in the United States. He reported the finding by the accident investigators of a poor arrangement of metres, dials and switches in the control room.

> In some cases, the distribution of displays and controls seemed almost haphazard. It was as if someone had taken a box of dials and switches, turned his back, thrown the whole thing at the board, and attached things wherever they landed. For instance, sometimes 10 to 15 feet separated controls from the displays that had to be monitored while the controls were being operated. Also, sometimes no displays were provided to present critical information. (Hopkins 1979 in Chapanis 1996, p. 1)

The list of user interface errors is a long one that includes missing or poorly worded labels; confusing menu choices; unclear, or an absence of, error messages; poorly constructed or missing online help; and crowded screens that cause difficulty with navigation. It can also include switches and levers in awkward positions, and issues with the layout of switches, meters, wiring diagrams and similar equipment. We look at these issues in Chapter 8 when we investigate the range of difficulties end users have with new tools and technologies.

DESIGN CONSTRAINTS

Designers work within a range of constraints that include budgetary and time limits, lack of sufficient resources and pressures created by manager and client expectations. All of these factors and more have a bearing on design error.

So far, we have looked at designers and the mistakes they and other members of the design team make that can create design errors and reduce the safety outcomes of new designs. But to be fair, we have to admit that design teams are not the only people who have control of design situations. Beyond design teams, we have the clients who request new or modified designs, design company managers and their decisions, politicians and political policies, banks and their control of financial capital, world stock market gyrations, international issues such as the global financial crisis (GFC) and the daily seesaw of world stock market pricing. All of these factors and more can impose external pressures on the design process in the form of constraints.

OPEN-CUT MINE RADAR SOLUTION

I spoke recently with the manager of a fast-growing business that designs and builds radar technology to inspect walls of open-cut mines so that operators can determine when a section of wall is unstable. A warning generated by this technology allows the mine staff to take remedial action to prevent cave-ins and rock falls, and avoid the accidents and deaths that could result.

His team was developing a new version of the technology with added safety features. Pressure was building from mine operators in several countries to have the new version released as quickly as possible. There had been delays, so the design and development team members were being hounded to get it out the door quickly. The pressures were building, not only on the developers and engineers, but also on a number of support staff, including the technical writers preparing user manuals and other support materials, and the translation people who had to convert these documents into several languages. Not only did the paper and online documentation have to be translated, but also the screen displays, button labels and menu choices had to be too. An added pressure was a new demand for the technology from Brazil, so a Portuguese-speaking technical translator had to be found in a hurry. None of these people could really get going until the technology was completed, tested and found to be solid and error-free. So these pressures placed extra constraints on the design team.

The manager suggested that the whole team should work longer hours and come to work on weekends to complete the project in a shorter time. I explained that the constraints being placed on the designers meant they would have to decide on some trade-offs. In many cases, I have witnessed, under a trade-off scenario, unit testing getting less than rigorous treatment, and a transparent user interface falling by the wayside. Developers spend less time commenting their code. This creates confusion and wasted time if that section of the project has to be reworked. If the deadline was not extended, I advised, they ran the risk of design errors slipping through that would almost certainly result in safety hazards in the mines, which could cause accidents and backfire on the company.

WALKING THE TIGHTROPE

Designers work within a climate where they have to walk a tightrope. They have to balance a range of factors and expectations, just like juggling balls and keeping them all in the air. In a recent study, designers expressed concern at the way a variety of constraints interfered with their ability to produce a safe design product. Designers singled out the pressures of limited finance, time, unrealistic client expectations and lack of appropriate resources as common constraints on their ability to produce safe and error-free designs. Another constraint mentioned by some writers was the need for designers to ensure the systems, devices and processes they developed were safe for end users and the public at large.

The designers who agreed to be interviewed in my study expressed concern with the need to trade off or balance time and effort between competing pressures to achieve an outcome. They saw this balancing action as a risk that opened the way for the occurrence of further design error.

Financial pressures were seen to pose the greatest risk to safe design, closely followed by time limitations, client expectations and a lack of appropriate resources. *Cultural conditions* took in the work climate and pressures from other external factors, and *policy* related to workplace business processes imposed by departmental or company managers. Other responses included the following items:

- Technology/production process limitations
- Internal politics
- Government regulations

UNSAFE ARMY VEHICLE

I spoke with a designer who, with his team, was working on the design of a new army vehicle. When pressed on the issue of client expectations, he made this comment:

> Sometimes we have to deal with stupidity. The risk manager was concerned that the soldiers travelling in the vehicle could be ambushed and they may have to get out quickly. He didn't want the risk of them bumping their heads against a structural member. That steel beam was essential to the strength of the structure to allow the vehicle to be collapsible, a major requirement.

'Don't they wear helmets?' I asked him. 'Of course', he replied, 'but he didn't want to listen to that argument'.

The pressures outlined above no doubt exist in any design business where managers, boards and shareholders are keen to minimise turnaround times and maximise profits. Designers expressed the view that design company leaders and corporate clients need to be convinced to view the design process of large, complex systems realistically.

***

Before examining the causes of design error in more detail, it will be useful to find out just who a typical designer is.

3 Who Is the Designer?

Our opportunity as designers is to learn how to handle the complexity … and make complicated things simple.

Tim Parsey

Just who is this person that sits in the middle of the design process? The designer can be a single person working alone or a team of people working in concert. This person or team works at satisfying a need by solving a problem with a new or modified solution.

Almost all designs begin with the need to solve a problem. Someone wants to build a new house, a company needs a new customer database or an engineer is dissatisfied with a tool or technology. Someone imagines a new system, device or process and sets out to build it.

Some years ago, a friend of mine built a small hovercraft. His design was a one-person vehicle that floated a small distance above the earth and was powered by a lawn mower engine. It was hard to steer, but his children had fun floating around the back garden of his house. It was an interesting prototype, and one he enjoyed building, but not worth pursuing any further.

More recently, companies are experimenting with driverless cars. They tell us that central city locations will see cars with no steering wheel, no accelerator and no brakes by 2020. Warren Buffet, who makes a lot of money from automobile insurance, believes the car insurance market will slowly disappear as self-propelled and directed vehicles refuse to run into each other.

Whenever anything is designed, no matter whether it is an information and communication technology (ICT) system or ICT-driven technology, an engineered device or a set of work procedures or business processes, a sequence of steps must be followed. This is true of all design. It does not matter whether we are designing a new magazine advertisement, a child's tree house, a cell phone app or a rocket to blast into space.

The first step requires us to create a design concept. This step usually involves gathering information and assembling it in a new way to form a concept that will be expressed through the new design. It may come as a 'light bulb flashing' moment, but most design requires deep thinking and problem-solving, and often results in a pile of discarded sketches or notes before the final solution is agreed.

HOW DOES A DESIGNER CREATE A NEW DESIGN?

My investigation into the ways designers reach a design concept showed, as you might expect, that they vary enormously in the methodologies they adopt. In order to understand the design process better, a study was made of designer behaviour and problem-solving methodologies. One of the topics looked at was intuition, on which writers offer varied opinions. We will also look at designing styles, problem-solving methods and conceptual strategies.

INTUITION

I believe in intuitions and inspirations....
I sometimes FEEL that I am right. I do not KNOW that I am.

Albert Einstein

In this study designers from around the world were asked to indicate their understanding of the term 'intuition'. 'Instinctive knowledge or belief' was the most popular choice, followed by 'a perceptive insight'. Fourteen per cent of the designers in the study indicated they do not use intuition in their design work.

Other definitions for intuition offered were

- 'Result of processing non-quantifiable and subjective information leading to a conclusion that can hardly be justified'
- 'Logical reasoning, perceptive insight, and creative contemplation'
- 'Knowledge and belief based upon extensive experience'

The word 'intuition', my dictionary tells me, comes from the Latin verb *intueri*, which is usually translated as 'to look inside' or 'to contemplate'. Intuition is thus often conceived as a kind of inner perception where the processes by which a 'knowing' occurs remain mostly unknown to the thinker, as opposed to knowledge that comes as a result of rational thinking.

My wife and I were discussing the meaning of intuition.

'Surely babies have intuition', she said. 'They know how to suckle at the breast almost from the moment of birth.'

'Is that intuition or is it instinctive behaviour?' I asked. 'Is there a difference?'

Animals are described as acting instinctively, but less so humans. Discussion in recent years rejects the concept of humans acting instinctively. It is argued that the human mind can over-ride instinctive behaviour through reasoning. But I would argue that a newly born baby is acting instinctively, the instinct to suck being built in atavistically, driven by an inbuilt desire to survive. I would argue further that having learnt that sucking at the breast provides satisfaction, the baby may then operate intuitively in seeking more food. I watched a young baby being held by a male. It opened its mouth and attempted to find a nipple to suck. It was obvious that the baby had learnt where to suck, but had not yet learnt that males cannot offer sustenance.

So intuition, it seems, relies on acquired knowledge. That knowledge may have been internalised a long time before, and relegated to long-term memory, but is still available when required.

Some years ago, I was working with fellow university staff as we built expert systems. We were experimenting with producing artificial intelligence in computers. Every interaction with a person was lodged in a database and could be called on when a key stimulus occurred. As a trite example, I taught my computer that when I was faced with a choice between jelly and ice cream, I preferred ice cream. The operation of this choice is shown in the following dialogue:

Restaurant waiter: For dessert you can have jelly or ice cream.
Computer: Ronald would prefer ice cream.

We could say that this example proves the computer was exercising intuition. It was making a choice between two possibilities based on learned knowledge.

A real-life designer can call on intuition to help solve much more complex problems. Let us say that a designer is working on a plan for a new bridge. He or she needs to consider a wide variety of factors, including the length, height and width; the weight that the bridge needs to carry; wind factors; maximum traffic density; and suitable construction materials. All of these factors and more need to be considered. The designer needs also to choose an appropriate style of bridge. Is it to be a cantilever bridge, a suspension bridge or take some other form?

The decisions that must be made will be based on knowledge about previous bridges this and other designers have designed and built. That knowledge base will possibly also contain information about bridge collapses. We could say that the huge database available allows the designer to rule in or out certain styles and characteristics of a bridge. As we now know, much of that decision-making can happen intuitively, that is without conscious analysis of all factors. That being the case, I would say all designers act intuitively at least to some degree.

DESIGNING STYLE

Designers were asked to choose a definition from a list that best reflected their designing style. 'Researched design' was chosen by the largest number of designers with 'user-focussed design' being almost as popular. This second choice seems to be in conflict with responses to other questions that suggest designers are rarely concerned with user-focussed design. This will be discussed later.

PROBLEM-SOLVING METHODS AND CONCEPTUAL STRATEGIES

The design process is a problem-solving domain, and it was deemed important to gather evidence about the conceptual strategies employed by designers. 'Research' was chosen by the greatest number of designers. This supports the earlier question regarding designing style where 'researched design' was chosen as the most popular designing style. 'Brainstorming' and 'divide and conquer' (i.e. breaking down a large, complex problem into smaller, solvable problems) were also popular responses.

Further questions designed to draw out a clearer definition of what is meant by 'research' made it obvious designers tend to be influenced by experience with similar problems, and often use an earlier solution to shape their response to the current design. As was hinted at in Chapter 2, their research often does not include calling on the experience of end users.

DESIGNER PROFILE

The study drew together the highest-ranked responses from those who responded to the designer survey into a profile of the average designer. This is not meant to denigrate any designer reading this book. It simply reports the responses from almost two-thirds of the designers from around the world who offered their experience and opinions in this study.

DEMOGRAPHICS

A majority of the designers who completed the survey were males aged over 36 years who had worked for 12 years or more as a designer. At that time, most of them lived and worked in Australia, the United States or the United Kingdom.

DESIGN ERROR OCCURRENCE

They agreed that design errors occurred in both the design process stages and associated business processes. When end users have problems using a new design, the designers were more likely to blame lack of training than inadequacies in the design of user interfaces.

SYSTEMS DEVELOPMENT LIFE CYCLE MODEL CHOSEN

On a recent project, the typical designer used a traditional systems development life cycle (SDLC) model where the end point is clearly identified before design begins and where there is little, if any, involvement by end users in the design process. The implications are that the design concept may not reflect the operational needs in the workplace and expensive reworking may be required after the design object is implemented.

CONCEPTUAL STRATEGIES

The average designer in this study said he may not use intuition in his design role, but does use a researched decision-making process. If a researched design approach means that designers carry out a thorough investigation of all aspects of the new design, the design concept should then be soundly based. If, on the other hand, it means relying on old solutions to new problems without investigating the current operational needs, then the design could fail.

CONCERNS

He is concerned about the constraints of insufficient budget, time and resources and client expectations, which can impinge on his work and open the way for errors to occur in the design.

TABLE 3.1
Designer Profile Items

Characteristic	Designers
SDLC model choice	Uses a traditional SDLC model that is more likely to be the V-model, evolutionary, prototyping or waterfall model, where the end point is known from the outset and end-user participation is minimal.
Intuition	May not believe he uses intuition in design work.
Design methodology	Researched design.
Problem-solving strategy	Uses a researched design approach to provide main problem-solving strategy, but may also use brainstorming to some extent.
Constraints	Is concerned about the constraints placed on his work by financial, time, client expectations, and resource pressures.
Human factors theory	Indicates he has had some exposure to human factors theory.
Stakeholder contact	On any given project, discusses the design with clients/stakeholders less than five times. More than three-quarters of stakeholder contact is with clients, i.e. business managers who order and pay for the design. Less than a quarter of all contact is with end users.
Blame focus	When errors with his design are attacked, he is more likely to place most of the blame on end-user human error or a lack of training.

USER-CENTRED DESIGN

He rarely speaks with end users and only communicates with clients (those who engage his services) a few times during a design. When his design is blamed for errors, he is likely to try to shift the blame onto end-user error and lack of training.

DESIGNER PROFILE ITEMS

The commonalities between designers were drawn from the responses to the questionnaire and interviews and formed into Table 3.1 for quick comparison.

***

Having learnt a little about who the designer is, we investigate the things he or she does during the design process in Chapter 4.

4 What Is the Design Process?

Recognizing the need is the primary condition for design.

Charles Eames

The steps one must take to develop a new design concept and work it through to maturity are described by the term 'systems development life cycle (SDLC) model'. These steps were first formalised for the information and communications technology (ICT) industry, but in recent years, formalised design process steps have been adopted and adapted by designers in other fields, including architects, engineers and business work process designers.

DESIGN CONCEPT FORMATION

All design stems from a problem or issue, so the first thing that must be done is to state the problem as clearly and succinctly as possible. All aspects of the problem need to be examined and every factor considered. We need to consider the five Ws – why, who, where, when, and how. To demonstrate how this works, let us look at the problem of accidents at rail level crossings in Australia. There were 74 deaths due to collisions between trains and motor vehicles between 1997 and 2002 (Australian Bureau of Statistics 2002) and many more injured travellers. Let us begin with a series of questions:

- Why? … Because of the large number of accidents, injuries and deaths at crossings that are not equipped with flashing lights or boom gates, because of the cost of repairing damage to railway and road vehicles, because of the delay to rail services, and because of the trauma to all involved, not forgetting for a moment the medical costs.
- Who? … Road travellers who, for whatever reason, do not look before driving across a level crossing, those who have a hearing problem and do not hear the sound of an approaching train, or those stupid enough to think they can beat the train to the crossing (and there are documented cases where this was given as the reason).
- Where? … There are thousands of unequipped level crossings in Australia and statistics to tell us which ones have the highest accident rates. Elimination must begin with the highest-risk sites. Unfortunately, these are often in rural settings a long way from population centres and normal services.

- When? ... As soon as possible. The speed with which a solution can be found and implemented depends on the amount of funding available and the political will to see a solution found quickly.
- How? ... This is where the designer must brainstorm possible solutions and evaluate between them by listing all the relevant factors, so that a cost-sensitive and operationally effective design can be devised. The decisions made at this time will be governed, at least to some extent, by the constraints imposed by the tyranny of distance. In remote areas, there will most probably be a lack of telecommunications coverage and a lack of electricity unless the rail service is electrified, although solar power might help here. Beyond that, there could be the cost of getting machinery and installers to the site.

Defining the problem requires detailed research. This may include looking at solutions to similar problems locally and abroad, talking to subject matter experts (SMEs) and investigating legislation that bears on the issue at hand. At this stage, it would be valuable to visit a worksite or location where the design solution will operate.

Once all team members agree they fully understand the problem and its operational climate, it is time to brainstorm possible solutions. It is important that when brainstorming, all suggestions should be recorded without judging against any criteria. Once all possible solutions are listed, it is time to agree on a set of criteria before evaluating the items on that list. Eventually, a solution will be agreed as the best way forward.

Now the agreed solution needs to be broken down into its component parts, each component to be a separate stand-alone portion of the whole. When all are agreed on the component list, it is time for the technical writers to begin drafting the specifications. The specifications and drawings need to describe each part of the design in detail.

WRITING SPECIFICATIONS

The design concept must be committed to specifications that describe each component of the system fully. It is crucially important that those who write or build the design understand exactly what is required. There must be no ambiguity, no room for misinterpretation. Testers will be validating the final product against these specifications, so they must describe the design concept accurately to allow the whole concept to be tested, as well as the component parts.

Specifications become the road maps to guide the developers, engineers and business personnel through the development phase. They must accurately capture every aspect of the design concept and answer questions such as

- Who is involved at every stage of the development?
- What resources are required to produce a safe and operationally successful design?
- What are the required outcomes of the new (or reworked) design?
- Where are the potential hazard pressure points in the design process?

Depending on the project, there can be a number of different specification documents written by the technical writers. The list can include some or all of the following:

- System specifications
- Individual component specifications
- User specifications
- User interface design guidelines
- Database design
- Data dictionary
- Developer style manual
- Models of required reports
- Error and incident logs
- Customer interface design guidelines
- User manual
- Assembly instructions
- Quick checklist of processes and procedures

People need to use these documents and drawings to build the design from those specifications, test the build and then install or distribute the new product or process so that others can use the new design easily.

BUILDING THE DESIGN

Depending on the type of design, someone (or a team) has to take the specifications and reflect the design concept in a system, tool, device or set of rules, processes or procedures. At this stage, it is important to refer constantly to the specifications and ensure they are being carried out accurately.

TESTING

The testing phase is as important as the formation of an accurate design concept and, like the concept stage, requires the participation of the whole stakeholder team. There are several types of testing and several ways this can be carried out.

A complex design is always composed of individual component parts. An ICT system has many separate operations, each one a stand-alone process. An engineered device will have independent parts. A business procedure will have separate rules, paragraphs, chapters and tasks. Testing these components separately is normally called unit testing and is often carried out by members of the development team, although each test should be repeated by the test team for verification against the specifications.

When all the parts are assembled into a whole project, the testing team will test the complete design against the specifications. Any anomalies will be sent back to the design team for review and reworking.

Testing teams often use technical testing tools where individual test cases can be constructed and applied to the design. For instance, an ICT system can be tested automatically for specific activities, such as adding a new customer to a database.

One of the most important activities is to involve end users in the testing activities because they will test the design, not against the specifications, but against the operational requirements for the tool or process. When this step is ignored, too often tools or processes are implemented that do not match the operational needs, thus opening up the potential for accidents to occur.

IMPLEMENTATION

It is vitally important to involve end users in the deployment stage. Too many reports indicate negative reception and lack of commitment when a new tool or process is implemented in a workplace without prior warning. On the other hand, when end users have participated in the design stages, the new technology is normally expected and welcomed.

TRAINING

Often, the need for training is overlooked or bypassed because of cost. The implementation of new tools, technologies and processes should always be accompanied by training. If team members have participated in the design stages, they make good ambassadors and can train their peers. Training can take many forms, from face-to-face sessions to videos, films, online webinars, and structured learning activities. I have worked with one company where all training is done online and successful completion of a course is recorded as points towards a salary increase.

MAINTAINING

The maintenance stage is often ignored. Once the new design is installed, many managers assume that the problem has been solved and no further action needs to be taken. But, nothing could be further from the truth. Many design errors only become evident when the new tool or process is put into use. This is when the anomalies, the misunderstanding of work processes, the poorly designed interfaces, the confusing controls, the mislabelled switches and levers, the controls that do the opposite to what was intended in the design and a profusion of other technological and human factors issues become evident. These are the things that can lead to incidents, accidents and fatalities.

It is important for the whole design team to recognise that the initial deployment should be regarded as a trial period, apprenticeship or internship. Design team members should be present to review the initial reactions in the workplace and be prepared to ensure that the problems that have come to light in initial operation are taken back to the developers, builders or writers so they can be reworked, and the anomalies and interface issues ironed out.

Misleading interfaces need to be adjusted; a confusing flow of processes needs to be corrected; changes in workplace tasks need to be updated by retrofitting systems, tools and procedures. Too often, accidents are caused when those maintenance activities are not done and end users have to devise workarounds because the tool does not allow for operational tasks to be done quickly, safely and effectively.

IF A DESIGN MEETS THE SPECIFICATIONS, CAN IT CONTAIN ERRORS?

A design can meet all the specifications and yet not perform in an error-free manner for the end user. This happens when neither designers nor clients consider the human factors issues surrounding the way an end user will work with the design in his or her environment. So many objects, technologies and business processes are designed with no thought given to the human factors issues. So many designers do not seem to consider the operational environment within which their design will operate, or the ways people will interact with the design product.

THE MANY FACES OF THE DESIGN PROCESS

We have seen that the design process consists of a number of steps. Those steps need to be followed no matter who the designer and what the methodology that person or that team adopts.

***

In Chapter 5, we explore the range of design process models.

5 Systems Development Life Cycle Models

There are many systems development life cycle (SDLC) models, some of them having been in use from the late 1960s. SDLCs were originally developed to guide the development of information and communication technology (ICT) systems. Today, they are appropriate for almost any project. In this book, I have limited their use to the development of ICT systems and ICT-related technologies, engineering projects and the development of business work processes and procedures.

SDLC models fall into two main groups, which, for clarity, we will call *traditional* and *adaptive*.

TRADITIONAL MODELS

A number of traditional SDLC models were developed beginning with the *waterfall* model that first appeared in the 1960s.

WATERFALL MODEL

The original SDLC, the waterfall model, was a method where a design passed through its different stages in order, something like water cascading down a waterfall. This model was first suggested in 1967 and used to guide ICT development. Later, it was borrowed by other designers, particularly engineers. It suffered from some serious defects, particularly the inflexibility created by not allowing reworking of aspects of the design during the design process, and not including participation by end users. The end point of the design was stated at the beginning and could not be amended during the process. When it was found that the design did not meet all operational requirements, the need to begin again meant blowouts in both budget and time.

A number of other traditional SDLCs evolved over time, each trying to improve on the waterfall model and correct some of its inadequacies. They were all marked by a fixed end point and either no consultation with end users or, at best, minimal contact.

The waterfall model has often been criticised as a linear model, a slow and heavy process that does not allow for correction of design errors before being deployed. Reworking is time wasting and costly, but it must be acknowledged that the waterfall model offers a structure that has been popular for many years and can still be found guiding system developers today. End users do not participate in the design and development stages, and they do not have the opportunity to evaluate and comment on the system until it is installed. There is no guarantee that the end product will meet the users' needs, and any serious problems in the design require a return to the very first stages of the life cycle, resulting in long delays and budget blowouts. Projects developed using the waterfall method can take years.

V-Model

The *V-model* life cycle diagram, like the waterfall model, describes a linear system where each phase must be completed before the next phase begins. The important change in this model is that every stage involves testing. The continual validation against specifications lessens the risk of errors and provides the opportunity for end users to participate in the development stages to some degree. The V-model became popular in Germany where it became the model required for all military, civil and federal information technology projects.

The V-model is considered to be rigid and inflexible, and has been found to be an expensive model compared with the waterfall model, needing more resources and extra money and time.

Incremental Model

This is a method of systems development where the model is designed, implemented and tested incrementally (a little more is added each time) until the product is finished. It involves both development and maintenance. The product is defined as finished when it satisfies all of its requirements. This model combines the elements of the waterfall model with the iterative feature of *prototyping* that will be investigated later.

The *incremental* model performs like the waterfall model, separated into overlapping sections. This model attempts to produce usable functionality earlier in the project life cycle. It divides the software to be developed into modules, which are then developed and tested in parallel. These modules or cycles are divided into smaller, easily managed iterations. Each iteration passes through the requirements, design, implementation and testing phases. In this model, the functionality is produced and delivered to the customer incrementally.

This model echoes some of the benefits of the V-model over the waterfall model in that smaller chunks of the system can be tested and installed for end users to validate before proceeding to the next chunk. There are some disadvantages of this system. Each phase of iteration is rigid and the iteration phases do not overlap, thus causing a disjointed product that lacks a seamless transition from stage to stage. It is also possible that problems can arise because not all requirements are gathered initially for the entire life cycle.

Evolutionary Model

May and Zimmer (1996) describe the *evolutionary development life cycle* (EVO) as a model that divides the development of a system into smaller, incremental waterfall models. Normally, the tasks are divided into small units that can be completed in a short time, perhaps a week or a fortnight. The major improvement of this model over the traditional *waterfall* model is that users get access to the product at the end of each cycle. By seeking feedback from end users at the end of each cycle, May and Zimmer see a significant reduction in risk from design faults.

Johansen and Gilb (2005) agree that seeking feedback from end users as the project develops lowers the risk profile. They also point to the earlier delivery of completed systems and the improvement in product quality. To demonstrate the quality improvement, they pointed to huge improvements over the operation of *Confirmit 8.0* (a software system built initially using the traditional waterfall development life cycle model) by *Confirmit 8.5*, which was developed using EVO. This system was designed as an internal reporting system in a large organisation.

Confirmit 8.5 took just 15 seconds to generate a survey instead of the 7200 seconds Confirmit 8.0 took. The time to set up a typical market survey was more than 3 times faster, and the time to grant a group of end users access to a report was 16 times faster. Surprisingly, the maximum number of simultaneous responders executing a survey jumped from 250 users for Confirmit 8.0 to an amazing 6000 users for Confirmit 8.5. So in performance terms, the evolutionary model was streets ahead of the waterfall model.

Johansen and Gilb agree the focus on early delivery of high value provided by seeking and using feedback from stakeholders is a positive aspect of the model. They approve of the attention shown to quality requirements and saw this as balancing the over-riding concern with functionality shown in other models. Negative aspects of the model include

- Defining requirements can be difficult.
- Finding practical ways to measure real product quality can be an issue.
- Delivery of a finished quality product can take longer than with other models.
- Testing can be postponed in the haste to get a module to stakeholders, and sometimes deferred testing is never done.

SPIRAL MODEL

The *spiral* model was defined by Boehm in 1986. This model was not the first model to discuss iterative development, but it was the first model to explain why iteration matters. The spiral model is similar to the incremental model, with more emphasis placed on risk analysis.

The spiral model has four phases: planning, risk analysis, engineering and evaluation. A project repeatedly passes through these phases in iterations (called spirals in this model). The baseline spiral begins in the planning phase, where requirements are gathered and risk is assessed. The subsequent spirals build on the baseline spiral.

This model combines features of the prototyping model and the waterfall model. It includes the twin advantages of breaking the project into manageable chunks and allowing end users to have access to and validate one part of the system at a time, in a manner similar to that of the V-model and incremental model. In addition, it includes a valuable risk assessment feature. The disadvantages are that management is more complex and the end of the project may not be known early.

PROTOTYPING

Prototyping models heralded a paradigm shift in SDLC models. The traditional models, the examples above being just a few, are clearly top-down models where the goal is clear; the end point is known at the time of beginning the design phase. The more flexible iterative models in this group allow for some tinkering around the edges, but the general direction of systems development is set in concrete before development begins.

Prototyping models can begin when a general direction is known, but not necessarily the form of the final product. In other words, incomplete versions of parts of the new system are developed and trialled by users.

The important difference is that end users can evaluate designers' proposals by trying them out. Through this process, end users can indicate requirements that developers have not considered.

Prototyping has several benefits:

- The designer can obtain feedback from users early in the project.
- System specifications can be modified before the final version is built.
- Initial project cost and time estimates can be more accurately measured to gauge whether original milestones can be met successfully.

Kay (2002) commented that a prototype can be created using tools different from those used for the final product, but prototyping is a 'front-heavy' process that can result in higher costs. He also indicated that this model can disallow later changes, an aspect that needs to be guarded against during later development.

Prototyping has been used within other models with some degree of success, the most well known being RAD, the rapid throwaway prototyping model and the evolutionary model, which we have already discussed.

ADAPTIVE MODELS

Around the year 2000, the first of what are known as the adaptive SDLCs was developed. The major differences are two. The first is that stakeholders and end users form an important part of the design team and are involved in regular meetings during the development. The second difference is that, because end users are involved, the end point of the design process is not acknowledged until the whole design team reaches agreement that the design will satisfy the requirements and solve the initial problem.

AGILE

Perhaps the most popular adaptive SDLC is the *Agile* system. At a groundbreaking meeting of 17 experts in 2001, the *Agile manifesto* was agreed:

- Individuals and interactions over processes and tools
- Working software over comprehensive documentation
- Customer collaboration over contract negotiation
- Responding to change over following a plan

They also agreed on 12 principles:

1. Our highest priority is to satisfy the customer through early and continuous delivery of valuable software.
2. Welcome changing requirements, even late in development. Agile processes harness change for the customer's competitive advantage.
3. Deliver working software frequently, from a couple of weeks to a couple of months, with a preference for a shorter timescale.
4. Business people and developers must work together daily throughout the project.
5. Build projects around motivated individuals. Give them the environment and support they need, and trust them to get the job done.
6. The most efficient and effective method of conveying information to and within a development team is face-to-face conversation.
7. Working software is the primary measure of progress.
8. Agile processes promote sustainable development. The sponsors, developers and users should be able to maintain a constant pace indefinitely.
9. Continuous attention to technical excellence and good design enhances agility.
10. Simplicity – the art of maximising the amount of work not done – is essential.
11. The best architectures, requirements and designs emerge from self-organising teams.
12. At regular intervals, the team reflects on how to become more effective, and then tunes and adjusts its behaviour accordingly.

The Agile SDLC model mandates that stakeholders, including end users, form part of the design team and meet regularly. In this system, work proceeds in short intervals, thus creating miniature projects that can be released independently. These short iterations may interfere with functionality. Real-time face-to-face communication is important, but it can leave a lot of documentation to be written post-project.

RAPID APPLICATION DEVELOPMENT

The *rapid application development* (RAD) model promotes a collaborative development environment. Stakeholders, including end users, are encouraged to participate in prototyping, writing test cases and performing unit testing. This model depends on a strong cohesive team that is prepared to participate fully in decision-making but not in engineering.

JOINT APPLICATION DESIGN

The *joint application design* (JAD) model uses a series of collaborative workshops where designers, developers and stakeholders work through the decision-making processes together.

Workshops are held over an intense two- to four-week period before development begins. Through these workshops, the stakeholders and ICT personnel can resolve difficulties and differences before development begins. Workshops follow a detailed agenda so that all issues are addressed.

Advantages include

- The time required for data collection, concept formation and specifications identification is greatly reduced.
- The sharing of ideas and joint decision-making help to develop a sense of project ownership and commitment by all parties.

Problems can emerge when

- The wrong participants are chosen for the workshops.
- Inadequate resources are available for the problem-solving activities.
- The facilitator either dominates decision-making or fails to draw timid participants into the discussions.

Some critics argue that involving clients and other stakeholders to the extent required by this method may create unworkable and unrealistic expectations.

Extreme Programming

In the *extreme programming* (XP) model, instead of detailed specifications being prepared, the programmers are asked to become totally familiar with what the end user needs. This process should be intense, immersive and thorough.

The focus is on getting the programmer to understand the end-user requirements and mindset, so that he or she can fill in the gaps the way the end user would want. This method stresses customer satisfaction and teamwork. Beck and Andres (2004) outlined elements of the system:

- Regular meetings with clients, commonly held weekly
- Minimised time-consuming reporting and other documentation
- Coding from day 1
- Progress governed by a clearly defined set of simple rules

Stephens (2008) argued that each rule depends on other rules, and if that co-dependence breaks down, the project will fail. He described the way constant revision of the design can allow the introduction of errors which may not be identified through unit testing.

Lean Software Development

In *Lean* programming, software defects are an accepted part of application development and eliminating them is a primary goal. This involves reducing the amount of code associated with error-free products, which in turn reduces inflated inventories and waste. For example, smaller pieces of pretested and error-free code are often used to build larger applications that meet customer needs. If properly implemented, *Lean* programming can deliver a complete product within budget and with greater efficiency, ultimately increasing customer satisfaction.

As with any methodology, the most difficult aspect of *Lean* programming can be convincing programmers to implement new development methods.

There are 10 simple rules of Lean programming:

1. Eliminate waste.
2. Minimise artefacts.
3. Satisfy all stakeholders.
4. Deliver as fast as possible.
5. Decide as late as possible.
6. Decide as low as possible.
7. Deploy comprehensive testing.
8. Learn by experimentation.
9. Measure business impact.
10. Optimise across organisations.

SCRUM

The *SCRUM* SDLC model, borrowing terminology from the game of rugby, improves productivity by allowing development to proceed in short bursts or sprints where progress is measured daily and is aided by constant communication between stakeholders and developers.

The *SCRUM* method embodies a 30-day cycle for delivering a working part of the system. Each sprint begins with a planning session that includes designers, developers and stakeholders, including end users. During development, a brief meeting of the team is held every day to keep the group focussed and to ensure progress is on target. This method is reliant on a 'SCRUM master', who has the ability to influence independent teams.

Tevell and Ahsberg (2011) believe that the practice of issuing regular small releases of parts of the system increases the overall quality and understanding of the system. It is believed that daily meetings lead to an improvement in the productivity of each of the team members, but the composition of the group is crucial to a successful outcome.

One of the potential disadvantages of SCRUM is 'project creep'. Because the end point is not determined before the project begins, there is a temptation for stakeholders to keep demanding new functionality. It is believed that this method works well on small projects, but is not suited to the development of large, complex systems.

***

As you might expect, there are many variations of the models explained above. Each design team needs to decide on a model that will guide the progress of their design process. In Chapter 6, we attempt to make that choice easier.

6 Choosing an SDLC ... Which Cap to Wear?

The design of information systems should be based on explicit analysis of work rather than assumptions about work.

K. J. Vicente

The systems development life cycle (SDLC) models outlined in Chapter 5 are only a few of the large list that exists, but they give a taste of the various approaches taken by designers around the world. One interesting model not included in this list was developed by Motorola in 1986. *Six Sigma* is not so much an SDLC as a set of quality management tools embodying statistical quality control to reduce costs by reducing defects in manufacturing.

Some of the design businesses I have worked with have developed their own slant on one of the models described above, but no matter what the model, the pressures on the design team are very similar. Managers want design products out the door as quickly as possible so that costs can be kept to a minimum. When safety breeches occur and those designs come back for reworking, no one is happy, yet so often the problems have occurred because of a flawed design concept caused by not consulting with end users, and this can have a falling domino effect on the rest of the design process stages.

The SDLC chosen for the development of new and the updating of existing technologies can have a strong bearing on the incidence of design errors. The choice of SDLC model affects the degree of participation by end users and other stakeholders in the design process, and impacts on the closeness of the fit of the new technology for the task that it sets out to support. In turn, this closeness of the fit of the new technology for the task governs the amount of retrofitting required. The amount of reworking directly affects budget blowouts and implementation delays.

Vicente, in his quote above, makes clear the need for designers to discover the true nature of the tasks for which they design systems. He and many others agree that the greater the degree of user participation in the design process, the greater the chance of achieving a fault-free system that closely fits requirements, that is delivered on time and on budget.

But the choice of SDLC is not an easy one, and depends very much on the design task. In broad terms, the design team needs to choose one of three options, depending on the design brief. Let us look at these choices, one at a time, and describe each in terms of wearing a different coloured cap. The first choice is the RED CAP option.

WHEN TO WEAR A RED CAP

There are many design briefs that are best developed by wearing a RED CAP. This class of project includes major works such as constructing bridges or high-rise buildings, sending a rocket to Mars or developing new government legislation. With this kind of project, the end point is set in concrete. The design must be thoroughly worked through with stakeholders and end users, and every aspect of the project signed off before the building begins. An *adaptive* SDLC is not suited to this kind of project. You would not want to build half a bridge and then have the end users wanting substantial changes. Perhaps cosmetic changes could be handled such as paint colour, but nothing structural. RED CAP projects need a *waterfall*-styled SDLC where the traditional design process is followed. There are a number of SDLCs that could be considered, from the *V-model* to the *incremental* model; even the *spiral* model might be suitable.

The important proviso that I recommend with this kind of design is that thorough discussion with all stakeholders and groups of end users must take place before the design concept is set in concrete. These people must have meaningful input into the design concept formation stage. Once they have signed off on the specifications, the design team should be left to get on with the construction. It would be wise to have an independent auditing team overseeing the construction to make sure the specifications are being followed to the letter. But if the design concept has been worked through thoroughly with the stakeholders and end users, there should be little room for design error.

It is a given that the design team will include specialists such as structural engineers, safety experts, human factors specialists and any other skilled people called for by the project. It is also a given that a detailed report on the site and operational statistics is available. For instance, if we were to plan the building of a bridge, we would need to know things such as the results of soil tests, predicted traffic flows and the required height above water at maximum flow rate.

A small bridge was built recently near my home in a Queensland country district. We were delighted to have a new steel and concrete bridge much higher than the old wooden bridge. The old bridge went under water every time a flood event occurred, preventing access to our local shops. In this sub-tropical climate, storms are common. Heavy rains caused a flood event a few weeks ago, and lo and behold, the new bridge was a metre under water, so we were still prevented from reaching our shopping precinct until the waters receded. We do not know who designed the bridge, but obviously the designers had not consulted locals who could advise them of the normal flood levels in this part of the country. If they had visited the site, they would have seen ample evidence of previous flood levels on the trees near the water course. 'Another design error', I said with a deep sigh. My long-suffering wife just smiled ruefully.

WHEN YOU CANNOT MEET THE END USERS

When designing something like a cell phone, it is impossible to consult with more than a small sample of end users. Companies are getting around that impediment by hiring user experience (UX) consultants. These people can be helpful, but they need to interact with as many potential end users of the product as possible, not just give their own impressions. In the best situation, these consultants will survey a significant sample of potential end users and relay their responses to the design team.

Some years ago, I worked for a time with the information and communication technology (ICT) team of a large government organisation. They were designing and developing Web software that would gather enrolment, assessment and examination data from all Year 11 and Year 12 students in the state. From those data, they would issue subject gradings and calculate a university entrance score.

The team I was working with was not skilled in user interface design, and I had serious concerns about the front-end design. I set up a series of workshops and called in staff from some of the schools to view the beta version of the system. As they worked through tasks I had devised, I recorded their problems and comments. My report was fed back to the design team and the issues I raised were dealt with before 'go-live'. This is just one of the strategies I expect UX consultants to adopt with modern systems.

WHEN TO WEAR A BLUE CAP

BLUE CAP projects are prototype projects. For instance, I have a brainstorm that suggests a new mousetrap, an improved cure for cancer or a modification to an agricultural machine. I need to follow my instincts by building a prototype that can be tested to see if my brainwave has validity. This kind of testing of a new idea is normally a one-person or small-team effort. Once the prototype has been proven to have legs, it needs to go to a formal design team and be built using an SDLC. Because this is a project with unknown factors and more than one possible outcome, it requires an *adaptive* approach using an SDLC like *Agile* or *SCRUM* or a modification of one of this family of SDLC models.

As we have seen earlier, adaptive SDLCs require involvement by stakeholders and end users right through the design process. Once the prototype has been proven and the project is formalised, we need to exchange our BLUE CAP for a YELLOW one.

WHEN TO WEAR A YELLOW CAP

Most ICT system design, engineering device development and business procedure writing are best served by wearing a YELLOW CAP. This means using an adaptive SDLC as the model to guide development. The Agile SDLC is the most widely known, but there are a number of other models in this family.

These adaptive models differ in two significant ways from *traditional* SDLCs. The first is that stakeholders and end users form part of the design team and are involved in every step of the design process. The second is that the end point is not exactly known until the stakeholders and end users say, 'Yes. This now suits our operational needs'.

It has been shown that adaptive methodologies get the project finished sooner and save money and time by not requiring retrofitting. They are also much more likely to have a low design error profile.

The downside is that, without clear ongoing communications and strong leadership, extra non-critical features can be requested at any stage of development. Progress through the design process needs to be smoothed by the leadership of a skilled negotiator, called a 'SCRUM master' in *SCRUM* SDLC terms. It is vitally important to reach complete agreement on the design concept and specifications before building begins through full and frank discussions with stakeholders and end users. With a clearly set-out beginning point and a skilled negotiator in charge, the problem of project creep should be avoided.

AVOIDING PROJECT CREEP

The adoption of an acceptance criteria approach has the advantage of avoiding project creep, a delaying mechanism than can develop when clients and end users continue to identify new requirements as the project proceeds. When acceptance criteria have been carefully crafted early in the project, new requirements can be assigned to later iterations of the design, and not allowed to delay the original deployment.

YELLOW CAP FUTURES: CONTINUOUS AGILE

To improve is to change; to be perfect is to change often.

Winston Churchill

One of the major issues facing designers is the need to manage the rapidly changing operational climate occurring with complex systems. The old 'build a system, device, process once' scenario no longer exists. The waterfall and similar traditional SDLCs are outmoded for most design projects. It is no longer possible to design and build a system once and expect it will serve its purpose for many years. Operational changes happen in workplaces almost on a daily basis, and designers need to come up with new ways of designing and maintaining systems so that changing circumstances in the workplace can be catered for quickly.

In 2012, an IBM white paper made the point that traditional SDLCs have become the problem rather than the solution for companies needing to improve the quality and time to market of new products. These out-of-time design processes are heavy and slow when the market really needs a rapid response to complex and ever-changing problems.

Agile and other adaptive methodologies that allow the constant involvement of all stakeholders in the design team and the iterative delivery of incremental value are the only SDLCs that can be adapted successfully to cope with constant change. For this purpose, I am advocating a methodology I have termed 'continuous Agile'.

Continuous Agile describes a new way of looking at the design process, and a new billing regime. We know from countless reports that teams using Agile methodologies can get their product to market sooner, cheaper and safer than teams using outdated traditional SDLCs. We just need to add a continuous maintenance billing model to the Agile model to allow the system, device or process design to remain open and ready to be modified at a moment's notice. Because the Agile SDLC already includes managers, end users and other stakeholders on the design team, they can reflect the constantly changing requirements accurately to the designers and test that the modifications asked for will meet the changed environment. A maintenance contract will no longer be an optional add-on. Rather, it will become a necessary part of the design billing structure. The original project team, or trained replacements, will need to be ready to reform and re-assess a design at a moment's notice.

A real challenge for designers is to ensure that ongoing modifications to one part of the design do not interact negatively on other parts. This safety-critical process needs very careful testing by competent and aware system testers who can view the whole range of interactions between components, and end users who can ensure the modifications deal accurately with the changing circumstances.

ORGANIC MODEL: A NEW LOOK AT THE DESIGN PROCESS

There is another way that SDLC models can be viewed. This view might give some readers a greater understanding of the design process.

In this approach, we apply the human experience to design concept formation and development. Just like us, each system has a birth and travels the path through early use or childhood to maturity before it begins to age and eventually reaches its use-by date or death. Just like us, it has an organic being. Within the system, components relate to each other just as our bodily organs relate to each other and play a part in assisting the actions of the whole.

It is worthwhile to remember the human analogy when viewing a design project. Not only the design activity product, but also the whole design process itself, has a life that is dependent on the involvement and sympathetic treatment by all stakeholders. If one part of the process, or design team, is out of kilter, then the process will be flawed and the end product of the design process will similarly be flawed. Unfortunately, the number of examples of flawed design in our modern world is overwhelming. Just some of those have been described in earlier chapters of this book. If you look around in your home, in your workplace, in the media or in government projects, you will find example after example of flawed design.

The challenge is to ensure the design process used to create new products, new tools and new processes does not contain flaws. And one way to do that is to develop an organic view of the design process where all factors are taken into account and where all stakeholders, including the end users, participate in the design process. This way, the elements of the process will be in balance and the whole will develop in a cohesive manner where every element is in sync with every other element and all will contribute to the common good. The model shown in Figure 6.1 portrays an organic view of the operation of any SDLC.

The outer ring of the diagram represents the stakeholders who have a vested interest in positive outcomes from the new technology. The inner ring represents the stages of development of a new technology or the modification of an existing system.

The *organic* SDLC requires managers and designers to come to terms with the participation of end users and other stakeholders in the design of the systems they will use. When end users participate in the development of the new technologies and processes, the designed product will reflect the role requirements more accurately, building acceptance and commitment on the part of the end users and, at the same

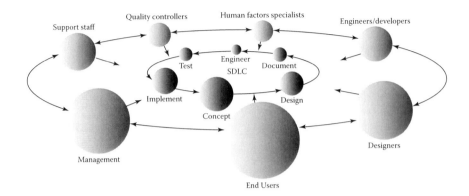

FIGURE 6.1   Organic SDLC model. Model devised by the author to show that design teams work within a much wider context that includes people such as support staff, stakeholders, management and end users. This wider view can be interpreted as an organic entity.

time, decreasing the chance of operational errors that could lead to accidents, avoiding retrofitting and saving implementation delays and extra costs.

The organic SDLC is an over-arching umbrella model. It takes into account all the factors leading to design error potential discovered in recent research. It operates in a cyclical manner with all stakeholders, including end users, represented on the design team. It is essential that end users form part of this team. They must participate fully because they know the operational climate within which the new design will operate. It is not sufficient just to consult with them. Results from consultation can be lost or ignored.

Where the end users are the public at large, the team needs to include UX specialists who can test every discussion point with a wide user base, and present as accurate a picture of user response as is possible at team meetings.

It is expected that the team will include managers both from the ICT, engineering or business design section or company and from the organisation requesting the new design. Other members of the team will include those who will build the design, technical writers, product testers, safety officers and other subject specialists who can bring outside knowledge and different ways of viewing the problem to the table, as well as end users who ultimately use the design product in their work.

***

Some rules that should be followed to produce an error-free design product are listed in Chapter 11. These rules are far more important that the SDLC model chosen and should take precedence over the format of the SDLC. The design rules outlined in my list, coupled with the organic SDLC, will produce an umbrella design environment most conducive to error-free, cost-effective and safe design.

Having looked at the various ways in which a design can take shape, we now need to examine where design errors can occur in a design process.

7 Where Do Design Errors Occur in the Design Process?

In Chapter 4, we discussed the seven stages in the design process. Now, we need to examine where things can go wrong as designers and other members of the design team work their way from the original design concept through to the final maintenance stage.

What constitutes the design process? There are five clear steps in any design process, and two more that should be included but are often ignored because of time pressures, lack of resources, overblown budget or a lack of understanding of the importance of these two steps. It does not matter whether you are wearing a RED, BLUE or YELLOW CAP, the following steps must be included in the design process. The systems development life cycle (SDLC) you are using and the type of design project you are engaged in will govern how much time and emphasis is given to each step, but each of these steps has relevance and, if not attended to carefully, will allow design errors to enter the process.

The five main steps are

1. Forming a design concept
2. Describing that concept in a set of specifications
3. Building, engineering and writing the design product
4. Testing the design build
5. Implementing the design in the workplace

The two steps that are often left out but should always be included in the design brief are

1. Training the end users
2. Maintaining the design product

DESIGN CONCEPT ISSUES

Trying to develop … products without gathering requirements first is planning to fail.

Robin F. Goldsmith

The most risk-prone design process stage is the very first step, the formulation of a design concept. The need for a design concept has arisen because of a perceived problem or issue that needs to be amended, or a new tool or process created. A design concept needs to take into account every factor, every variable associated with a successful solving of the problem. When a design concept is flawed, the rest of the design process is doomed to failure, so the challenge is to arrive at a concept that will lead to a successful resolution.

Sadly, many design concepts are flawed and either the resulting design fails to solve all aspects of the original problem or, worse still, design errors occur during the design process that can lead to accidents and deaths.

One of the more common reasons for a flawed design concept is a lack of rigorous research. Designers far too often make assumptions about what is needed without careful analysis and consultation with end users. The Queensland Health payroll fiasco described in Chapter 1 provides a clear example of the design team not communicating with the users to discover the extent of the problem that needed to be addressed. They used a previous system built for 1,800 workers on two salary scales to build a payroll system for 85,000 workers on more than 24,000 different pay scales. It failed dismally.

This lack of contact with end users often means designers lack a clear view of the operational setting and their designs are not as suited to the task as they might be. More than once, I have sent designers and developers to customer sites to demonstrate a new system or process. They have always returned with a much clearer understanding of the operational setting and the needs of the end users. After these visits, I have found them more willing to modify interfaces and processes for the people who will use this new design.

Another problem is concerned with the lack of group consensus among design team members when formulating a concept for the design. Sometimes, the highest paid person's opinion (HIPPO) dominates discussion and decision-making, and that is not always a good thing. Alternatively, the design team may be swayed by the client's expectations, and sometimes the client knows no more about the operational requirements than the design team.

So in brief, we can list less than adequate research, or reliance on faulty research, that fails to inform the team of the real requirements of the design. To this, we can add the confusing and sometimes unhelpful guidance from managers, or strong team members, and the often equally unhelpful information from clients.

The *adaptive* SDLC models, which require stakeholders, including end users, to be part of the design team, help to mitigate the above design concept risks. These people have an intimate knowledge of the workplace needs and are able to help designers get the design concept right the first time. Many teams these days adopt *Agile* or *SCRUM* SDLCs to give them the knowledge they need at the beginning of the project.

As systems, devices and processes become more complex, the design concept becomes even more difficult to get right. We look more closely at the issues related to complex systems in Chapter 9.

WRITING SPECIFICATIONS

Once agreed by all members of the design team, the design concept must be described in specifications that describe each component of the system fully, but no matter how meticulously the technical writers approach this task, if the design concept is flawed by not identifying and dealing with every variable, the specifications necessarily cannot describe a perfect solution.

Because the specifications may be flawed to the same extent as the concept, the build will also reflect that risk and the testing, which tests the build against the specifications, will also not pick up the flaw.

A perfect design concept cannot on its own guarantee a perfect design solution. Each of the subsequent design process stages has its own risks. The specifications spell out the fine detail of the concept. These are the instructions that others will rely on to turn the design concept into reality. If there are gaps, the way is open for developers, builders, engineers or writers to make assumptions about what the specifications have not said. It immediately becomes obvious that wrong assumptions by the builders will lead to built-in risks or hazards that can cause accidents in the future if not corrected.

A classic example is provided by Brisbane Rail's partly automated control system. In 2009 on three separate Friday evenings, the control system failed and around 100 trains were stopped (Halcrow 2010). Thousands of city workers were using those trains to return home. The railway company did what it could to organise buses to move as many of these travellers as possible, but on each occasion it took several hours before the jam could be freed and train services returned to normal. Luckily, no one was injured, but many were angry and frustrated by the delay. An information and communication technology (ICT) manager admitted to me, 'By mistake we developed that component to take in a text file instead of a binary file, and that may have caused the MR [message redirector] to fall over'. That mistake should not have happened if the specifications for the system had indicated clearly the kind of file that should be passed to the message redirector and the tester had validated the file type stated in those specifications was being passed.

Other mistakes can be made if the specifications do not include a data dictionary, a list of the words and their meanings that are operation specific. Everyone on the design team must be able to comprehend exactly what is intended. Confusion over terminology inevitably leads to misunderstandings and mistakes.

It is also very important to include a developer or builder style guide to produce consistency across the design team. Just as an author uses a publisher's style guide to produce consistency in spelling and grammatical expression, so too designers need to conform to a consistent pattern in coding and commenting on code, engineering terminology and business process language where a stylistic imbalance can create misunderstandings and result in design faults in the product.

Lastly, it is vital that the specifications include a style manual for user interface design. There are so many places in new designs where a poor interface description or indeed no interface information can lead to mistakes such as switches with no labels, as we saw with the Three-Mile Island nuclear disaster example in Chapter 2.

BUILDING THE DESIGN

The kinds of things that can go wrong when building a design are many. Some errors are caused by incomplete specifications where there is an insufficient explanation of the finer detail of the design, or gaps left in the description of process steps that leave it to the developers or builders to assume what is needed. Similarly, the lack of a data dictionary can lead to confusion over the meaning of specific terms, and the lack of a developer or builder style guide and the lack of user interface information can also lead to inconsistencies in approach and potential errors being built into the design.

But, there can be an even more sinister issue. Design builders normally work through some kind of job system that is passed around the team. This system contains details of every aspect of the build, who is responsible for each item, who has carried out a unit test on that item, the results of that test and what remedial action, if any, needs to be taken. In this way, every small detail of the build is tracked and signed off when it has been through all the required steps. When everyone on the team adheres to this business process and records the completion of each assigned task, the chances of errors creeping into the build are small. But for this kind of job recording tool to work effectively, each member of the team must use it religiously and someone, normally a project manager, needs to keep his or her fingers on the pulse and not let anything slip through the safety net. Unfortunately, I have worked on teams where only token recognition has been given to the value of this business process tool, and errors have slipped through the net. Sometimes, they are picked up by the testing team, but not always. This business process tool is incredibly valuable, and when used correctly, it can go a long way to helping to produce a safe design product. When only given token observance, it can act like a leaking sieve, allowing design errors to slip through the system.

But, there are other factors that can influence the build of a safe and error-free design. It is a commercial reality that most new designs need to be built and implemented quickly and the pressures of time to market, budget and availability of suitable resources can operate as constraints on the design process. When the pressure to complete a design quickly is intense, constraints demand shortcuts and attention to core functionality, as opposed to what seem less important features. One of the first features to be sidetracked is often the user interface. Along with it often go training and maintenance. Testing too can take a hit. If stakeholders and end users have not been included in the design team and allowed to influence the direction of the build, the way is open for design-induced errors at the workplace.

Good project management can help to alleviate these issues by accurately forecasting and balancing out the pressure times, and building in enough slack to allow the design to be built in the best possible way, but this does not always happen.

TESTING

Regular testing is done against the specifications. The aim is to prove that the design meets the concept expressed in the specifications. As you will understand, if the concept is flawed, the specifications will necessarily reflect that problem and the testing, which tests against the specifications, will not identify design errors that relate to flaws in the original design concept. The testing should of course identify errors introduced during the build.

When end users are added to the test team, they will be expected to test the design against the workplace requirements, to ensure the new tool will perform the required operational tasks correctly. It is expected they will identify errors in interpretation of the workplace process flow and user interface issues that could confuse or mislead other end users. There are many cases where further testing by end users should be done after installation of the new tool, technology or process.

I have described what should happen in a perfect design process, but unfortunately, too often things go wrong because of time pressures or staff and budget limitations. I have seen people who are not fitted to the task pressured into testing because appropriately trained staff members were not available. These people may not have an adequate knowledge of testing procedures. Some rely solely on automated testing tools. Some have no knowledge of the testing tools available. I have seen people testing the running of code in an ICT system but failing to check that what the code intends is reflected accurately in database records. The specifications can be incomplete and fail to include fine detail that needs to be tested. Often, there are no end users on the test team who can validate the design against the operational requirements.

Designs need two types of testing. There should be unit testing of small chunks of the design output, and there should be a thorough end-to-end testing of the whole design. One of the serious issues with the failed Queensland Health payroll system was ignoring the need for an end-to-end whole system test. Time was short and a whole system test was never carried out, with disastrous results.

IMPLEMENTATION

Implementation takes many forms. ICT systems can be placed on the Web for ICT support staff or users to download, or they can be delivered on disc or transferred in from a master server. Engineered devices may be bought off the shelf, arrive by courier, or be brought by technicians who will install them. Business processes may be lodged on a noticeboard, introduced during a staff meeting or emailed from a manager.

The delivery method is far less important that the human factors issues. Too often, the change comes as a shock because end users have not been forewarned. When new systems, technologies or processes arrive unannounced, you can expect them to meet resistance, negativity and lack of commitment.

End users can jeopardise a new program if they have not been prepared. There was a case not so long ago in Australia where a rail company introduced a new business process for train drivers to follow as a method of lowering fuel costs. Because they had not been consulted, the drivers refused to follow the new process, and they could not be shifted. The design of the process was flawed because end users had not been included on the design team. They held their line, with support from their union, and no threats or cajoling could shift them.

Successful implementations are those where the designers have included end users on the design team and have encouraged them to contribute to the discussions. In this way, they build acceptance and allegiance to the new design and are the best envoys of the change with their fellow workers.

TRAINING

Training is often ignored by managers or given only token observance. It is often assumed that the end users know their job and they will soon work out how to use the new tool or process. Others see the cost of training as an unnecessary expense.

I spoke with a manager of a safety-critical high-voltage power control operation. He talked about the high cost of training for the complex control technologies being installed. He told me that a training company had been employed to train all the workers in the first of these new technologies. The cost of training course development and face-to-face training had been exorbitant, and the company just could not afford to do that again. He asked me to develop some courses that could be delivered online so the staff could work their way through them in their spare time. Initially, there was some resistance to learning online instead of in face-to-face sessions, so it was agreed to link the successful completion of these courses to salary credits, a strategy that was well received.

Training cannot be ignored, particularly for safety-critical situations. In recent research with end users, I was told over and over again of instances where management refused to offer training. The following is just one of the perceptive comments made:

> Training needs resources and most decision makers perceive it as a cost, not an investment.

MAINTAINING

In my experience, far too many designs come without a maintenance contract. It is assumed that the system, device or process is fine like it is and, once in operation, needs no maintenance. Sadly, that is not the way things are most of the time. In almost all designs where end users have not been involved, there are problems.

When the need for ongoing maintenance is ignored, designers and often managers will not listen to the concerns raised by end users. They ignore calls for changes to improve safety and efficient operation, and do not understand, or want to know about, the problems with user interfaces. They seem oblivious to the potential for accidents when interfaces are poorly designed.

On one occasion, I was working with some Airservices controllers who control the airspace for half of Australia and surrounding oceans. Their operations were highly safety critical, and they used complex observation and control tools. A controller was complaining about one of these computer systems:

> Again it's one of those things where the end user doesn't have any input and there is no maintenance. This system was difficult to use so we had to figure out our own workarounds. Nobody asked us what we needed. Now they are going to update it and we'll have to start again.

PARADISE LOST

The seven steps outlined in the beginning of the chapter are not always dealt with as they should be, so an important item on the list of potential crisis points is the error of omission caused by ignoring the need to include training and maintenance in the design process. Designers can build the best of new technologies or work processes, but if the end users are denied training, as happens far too often, there is a serious risk of them not fully understanding this new tool or process.

Misunderstanding can lead to mistakes, and mistakes can lead to accidents. Lack of training is bad enough, but when it is combined with a lack of maintenance, the hazard potential can reach the serious stage.

Of course, if end users had been included on the design team, they would have been able to represent the operational requirements and the end-user perception of the user interfaces to ensure that those interfaces were intelligible and work process flows suited to the purpose. By being members of the design team, these end users would have been thoroughly familiar with the new technology, and able to act as peer tutors to their workmates. Furthermore, by being included on the test team, they would have been able to test the system against work demands and obviate the need for immediate maintenance.

The big benefits of this approach are positive attitudes and immediate acceptance among the end-user team. Negative attitudes are common when new technologies and processes are implemented without appropriate warning or training. We can expect fault finding and negativity when a new process or tool is imposed by management. When members of their team form part of the design team and are able to reflect user needs and perceptions, the rest of the end-user team are much more likely to accept the new tool readily. Sadly, the positive payoffs from including end users in the design team are not recognised often enough, and a valuable efficiency strategy is lost.

***

Many of the hazards described in this chapter touch on human factors concerns. The next chapter examines these and other HF issues in detail. My disconnect model that you will meet in Chapter 8 demonstrates one of the major HF issues, the lack of end user participation in design.

8 Human Factors Issues

Socrates said, 'Know thyself'.
I say, 'Know thy users. And guess what? They don't think like you do'.

Joshua Brewer

The results of my recent study into the design process indicated a serious disconnect between designer, client and end user. This disconnect explains some of the design flaws that can cause unsafe and sometimes disastrous operational outcomes. Designers indicated that more than three-quarters of their contact with stakeholders was with those people they describe as 'clients', the people who commission them and pay them, whereas less than a quarter of their contact was with end users. The clients in large organisations are often as far removed from the workface as the designers, yet they are the ones drawing up the design requirements.

End users from large organisations reported that they have little direct contact with the clients, the business managers who make decisions about the design and implementation of technological tools and new business processes. The designer–client–end-user disconnect model (Figure 8.1) describes that issue.

The disconnect model indicates that designer contact with clients is much greater than their interaction with end users. Results indicate a substantial gap between the formal user requirement statements from clients and the unvoiced end users' expectations. Designers' perceptions are consequently biased towards a problem definition that may not reflect true operational requirements.

Figure 8.1 also indicates some of the range of constraints, which include time, budget and resource limitations, as well as unrealistic expectations from clients. Those constraints militate against a meaningful level of discussion with end users where clear operational guidelines could be agreed. Much is written about the benefits of participative design where end users form part of the stakeholder team, but the results of this study indicate that participative design happened in less than a quarter of design projects.

One of the more aware designer participants made this comment:

The best strategy appears to be to involve end users in the conceptual design process. If you don't have a clear understanding of the users' needs to begin with, your concept is a lost cause even if it is innovative. Not meeting user needs and expectations is the most glaring error I've witnessed.

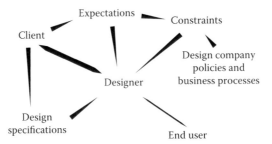

FIGURE 8.1 Designer–client–end-user disconnect model. Model devised by the author to show the major interactions and constraints operating on designers, each one of which can introduce errors into a design. It also indicates the lack of communication between designers and end users which leads to one of the major design fault causes.

USER INTERFACE DESIGN

One of the important human factors issues is user interface design. In a recent study, end users were asked to list the interface problems they had with new technologies. Those problems were divided into three separate groups, computer interface errors, non-computer errors and associated business process errors.

COMPUTER INTERFACE ERRORS

In the study, the most commonly voiced end-user concern with computer-based systems was the lack of error messages or, at very best, unclear error messages when operations go wrong. For example, if you were an end user faced with this error message, what do you think it might mean: 'The IO Thread was interrupted'? Perhaps it might help a little to see an alternate message, such as 'Your file [filename] could not be uploaded because the server [servername] closed the connection'. This would at least give you something concrete to take to the network techs.

The main problem with error messages is that they are written by the developer, who has an intimate knowledge of the system. This person often has little ability to see the message from the end user's point of view. He or she understands clearly what is wrong, but the end user is totally mystified and needs to seek assistance from a technical person to interpret the message.

The next problem voiced by end-user participants was confusing screen menu choices. One of the features rarely seen in complex systems is menus arranged according to the processes end users need to go through to carry out a task. The following example is fictional but representative of so many confusing interfaces.

The end user needs to add a new customer to the database. The steps in the process are

1. Open the database – Menu 1 item 3
2. Choose the form to enter a new customer – Menu 5 item 7
3. Allocate a purchasing authority – Menu 4 item 2
4. Validate the database entry – Menu 6 item 8

I am sure you have seen something like this in a software package. Now consider (1) the number of disjointed steps to learn and (2) the time this operation will take. Surely, a process-oriented interface should be able to put all these steps together in order and in one menu. Similarly, when documenters are writing user manuals, good communicators understand the way users need process steps, not disjointed menu item steps, to follow. So simple, but so often ignored.

Difficulty navigating around the system also rated as a high concern for many end users. The process view also applies to navigating through a system. The difference between this item and the previous one relating to menus is that with the end user having made one choice on the front page of a system, a second page opens and activities there lead to other pages without returning to the home page. Some systems handle this well, but I have worked with controllers in safety-critical industries where the progress through a process is difficult to follow and wrong turns can easily be taken. There are industries and times in any industry where accuracy and speed are of the essence, and the consequences of making a wrong choice when navigating through a system could be disastrous.

Lack of online help was also indicated as having a high priority for many end users. When end users ask for help while carrying out a process, they expect to be offered meaningful advice. Often, that online help is non-existent or, if it does exist, is less than clear and helpful to users.

Unclear screen design was also mentioned. Often, screens I have evaluated for clients have been jumbled with no clear order, have been crowded with a variety of typefaces and fonts or have used colour indiscriminately and badly. There are a few simple rules that have been determined over centuries and proven by modern research. The first is that in Western countries where everyone has learnt to read from left to right and proceed down the page one line at a time, computer displays should mimic this procedure, and not expect the viewer's eyes to jump around the screen. Because computer screens are of lower resolution than the printed page, and we recognise letters and words by deciphering their edges, it is better to use sans serif typefaces like Ariel or Verdana, rather than seriffed typefaces such as Times New Roman. For variety between headers and text, use no more than two typefaces on any page. The use of colour is fraught with difficulty. For clear reading, the closer the colours are to black on white or white on black, the better. Choices like yellow on red are difficult for the normal reader, let alone for those who suffer from colour blindness.

A badly worded or missing label is another problem for users. Every button should be clearly labelled so that users understand when it should be used. Linked to this is the need for process names on buttons to be the usual names used within the group. I have been mystified when a button or menu item uses a name that is not common for that process and does not describe the process in the way I would expect.

Poor testing can lead to unexplained wrong actions from key presses or no action at all. All key presses should be validated before the system is deployed. Serious accidents and deaths have occurred when key presses have not been validated in medical equipment. Steven Casey's account (1998) of design errors in the Therac-25 radiation therapy describes this problem clearly. Read the story in his book *Set Phasers on Stun*. He explains that a certain set of keyboard presses failed every time it was used, resulting in patients receiving lethal doses of radiation.

Non-Computer Errors

End users responding to the study indicated a number of problems not related directly to the user interface of computer systems. The most frequently indicated problem was concerned with controls placed in awkward locations. Those controls include toggle switches and dials, graduated controls and activity readout screens, gate valves, brakes and other levers, and foot controls as well. The Three-Mile Island nuclear disaster was caused, according to the accident investigators, by controls being placed well away from the dials and screens the operators needed to read while modifying settings.

End users from control rooms and towers complained about the problems they have when radio or phone communications break down. This becomes a serious issue when dealing with safety-critical activities. Some organisations I have visited have installed a second backup communications network that can be called on when the main system fails.

Many systems include alarms that are activated when a fault occurs in a system, either locally or remotely. End users complained that alarms for different faults were sometimes too similar. Others mentioned that alarms can be too noisy or too quiet. One would expect that the settings for these alarms could always be modified at the site, but in some cases, it seems the settings were hardwired in during the design build. If this is really the case, it constitutes a design error with potentially serious consequences.

Associated Business Process Errors

End users also mentioned some associated business processes that made their work more difficult. The first of these was unclear wording in process documents that could lead to misinterpretation. Another was the lack of a business process manual to help new workers become aware of the local ways of doing certain tasks. These customary behaviours are often unstated but understood by the people who have been in the area for a time. Associated with this was the phenomenon found in many businesses where some individuals assume responsibility for a specific set of tasks, but because of the lack of a written record, new workers are confused about who they should see about this or that. In addition, if that person is absent, questions only he or she can answer have to wait.

END-USER PROFILE

The study drew together the highest-ranked responses from those who responded to the end-user survey into a profile of the average user of designed systems, engineered devices or business processes. This is not meant to denigrate any end user reading this book. It simply reports the responses from almost two-thirds of the end users from around the world who offered their experience and opinions in this study.

Demographics

A majority of the end users who completed the survey were males aged 36 years or more who had worked for nine years or more in this role. At that time, most of them lived and worked in Australia, the United States or the United Kingdom.

Design Error Occurrence

They agreed that design errors often originate from design and associated business processes. In contrast to designers who are more inclined to blame a lack of training for end-user problems with interface design, end users placed more blame on the poor interface design of the technologies they use.

Problems Experienced

The average end user has experienced interface errors in technologies he uses, including unclear error messages, confusing screen menu choices, difficulty with navigation, lack of online help and unclear screen design. This person is also likely to have experienced controls placed in awkward positions and had problems with phone or radio communications. Also mentioned were the unhelpful business processes developed by people who did not understand the operational climate.

Participative Design

This person was not consulted by designers during the design of a recent tool he will use but believes he should have been. He believes he should participate in discussions during the concept development stage, testing of a new system and deployment or implementation stage.

Much is written about the benefits of participative design where end users form part of the stakeholder team, but the results of this study indicate that participative design happens in less than a quarter of design projects concerned with the development of new technologies.

Concerns

End users are likely to experience errors with equipment, and receive insufficient training on new technologies. The typical end user has not been consulted by the designer of a new technology, tool or process he will have to use, but believes he should participate in the design process at least to the extent of being involved with concept formation, testing of the new tool and implementation into the workplace. When errors with his tools occur, he is more likely to place most of the blame on design error or technical error.

COMPARISON OF DESIGNER AND END-USER PROFILES

Designer and end-user profiles were constructed from the data, to provide a focus for error causes and error minimisation strategies. The first six categories were common to both groups and displayed a degree of agreement between them (Table 8.1).

The results display a majority of males in both groups aged over 36 years, with designers having worked a few more years in the role. Both groups see design errors occurring in both the design process and associated business processes. Designers are more inclined to blame a lack of training for end-user problems with interface

TABLE 8.1
Comparison of Designer and End-User Profiles

Characteristic	Designers	End Users
Gender	Male	Male
Age	36 years or more	36 years or more
Country	Resident in Australia, United States or United Kingdom	Resident in Australia, United States or United Kingdom
Time in role	12 years or more	9 years or more
Believes design errors originate from	Design process and associated business processes	Design process and associated business processes
Interface error experiences	Believes that end users have most difficulty with system navigation and puts this down to a problem with training	Has experienced user interface errors in technologies he uses, including unclear error messages, confusing screen menu choices, difficulty with navigation, lack of online help and unclear screen design; he may also have experienced poorly located switches, levers and other monitoring and control equipment

design, while end users place more blame on the poor interface design of the technologies they use.

***

It is now time to investigate the introduction of automation (Chapter 9), the fast-moving trend that is taking over our world and having an immediate impact on our lives.

9 Automation ... Persistence of a Myth

Tesler's Law of Irreducible Complexity
When we add automation to simplify the demands upon people, we increase the complexity of the underlying technology. The more complex the underlying technology, the more opportunities for failure.

D. A. Norman
Living with Complexity, 2011

Dekker (2011) describes the drift into failure when mechanical or electronic equipment begins to exhibit stress and fatigue that is not identified by normal maintenance procedures. When a part in a plane, train or control system fails, it can come as a complete surprise, and often the failure is blamed on human operator error when really it should be regarded as design error or maintenance error. We could blame human error by designers, or human error by maintenance crew, but in reality, the end user, the operator, is often the first to be blamed instead of the last. This person is a convenient scapegoat.

HUMAN COST OF AUTOMATION

Most network control rooms I have worked in or visited are busy places with controllers actively monitoring network activity, communicating with workers in the field, making decisions about re-routing traffic and dealing with issues and problems as they occur. But one Australian railway network uses a new highly automated network control system that needs minimal intervention from the controllers. Their role is little more than a watching brief. They have become what Norman (2007) calls 'caretakers'. They have almost nothing to do until an emergency occurs. When that happens, they must act quickly. Their ability to remain in a state of high alert is seriously jeopardised by not having a more active role in the network activity. One of the managers responsible for this highly automated network control system commented during an interview, 'Our system got cleverer than the operators, and they had no idea what it was going to do. So we dumbed it down, we pulled a lot of cleverness out of our code'.

One of the controllers made the comment, 'The system is now working so well that network control is almost totally automatic and we don't have enough to do except when things go wrong. Maintaining heightened awareness and combating fatigue are difficult problems for us'.

But things did go wrong when a developer doing a regular upgrade uploaded the wrong files. The system stopped working, and so did all the trains. This time, it was human error at the systems end, not with the end users. The response within the

57

information and communication technology (ICT) section was to increase the rules and supervision concerned with downloading files to the network.

The normal response to what is perceived as human error is to tighten up procedures, increase rules and set up barriers so that operators become limited in the choices they can make. Each new layer of defence increases the complexity. This complexity, by its very nature, increases the chance of error. Complex systems are less safe. Operators become more frustrated and their fatigue levels increase. Norman mentions the problem of 'overautomation', where the system is so good the users do not have to pay close attention. There is little for them to do. Consequently, they find it difficult to respond when something does go wrong. Their ability to maintain heightened awareness is compromised.

COMPLEXITY

Figure 9.1 describes the growth in complexity and accident potential in the design of systems, devices and procedures.

At the bottom, we see the stock standard traditional type of design process. A certain amount of risk is involved, but if the design team sticks to a formal process with adequate checking and testing as they go, the complexity and risk potential remain reasonably low.

Layer 2 takes into account human factors both within the design team and externally with stakeholders and end users. When these groups operate independently with little or no communication or cooperation, the added layer of complexity increases the risk factors.

Layer 3 considers the pressures put on design teams by organisations that include both the design company or department and the client group. The typical triangular tugging between the demands of managing finances, resources and time results in trade-offs that increase the complexity of the design process and add further risk of error.

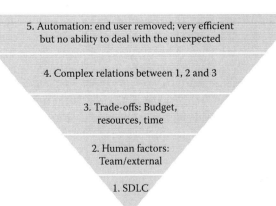

FIGURE 9.1 Five levels of complexity model. Model devised by the author to stress the greatly increasing complexity of systems as more levels of automation and thus complexity are built in. As complexity increases so does the risk of design error and the chance of accidents with far greater consequences.

Layer 4 shows that when each of the factors in Layers 1–3 is considered independently, the risk is predictable and measurable, but when we consider the interplays between these groups and activities, the level of complexity increases exponentially. Suddenly, the level of risk becomes almost too great to predict or measure.

Layer 5 describes the implementation of automation, particularly when end users are removed and machines are required to take on a much wider range of activities and responsibilities. Designing into a machine and its associated processes and procedures the full ability not only to do its assigned tasks, but also to mimic the tasks, thinking processes and decisions of the human it has replaced, increases the level of complexity to an almost immeasurable level. When accidents happen because of flaws in the design process, they will be much more serious and have far greater consequences in terms of governance, money and human life.

COMPLEXITY IN ACTION

In many countries, automatically controlled trains are required to have a 'driver'. This person maintains high visibility to reassure the travelling public and sometimes opens and closes the doors, or applies the brakes when required. Like the network controllers described above, this person does not actively control the train except when an unforeseen event occurs. The common complaint from these people is the difficulty of maintaining the high awareness needed to respond to emergencies.

Because the operator is seen, falsely in my opinion, as the weak link in the chain, systems are becoming more complex so that the operator can be removed. The number of remotely operated trains is increasing rapidly around the world. Currently (2015), there are 206 automatic rail systems around the world, with another 33 systems in the process of being built or converted. Train drivers are being replaced by electronic systems that are controlled from network control centres. This does not remove accidents. It just makes them more spectacular.

The disastrous high-speed rail crash near Wenzhou, China (2011), killed around 40 people and injured more than 200. Lightning struck one train and wiped out its electronic equipment, forcing it to come to a halt on the track. A second high-speed train ploughed into it, causing a serious rail disaster. Initially, the accident was blamed on human error, but it was eventually found to be caused by design error in the rail network control technologies (The Hindu 2011). The controlling systems were not sophisticated enough to 'see' the stopped train, and the controllers were unaware of the situation, so they did not warn the following train to slow down and stop. The well-designed Maglev track and trains were not matched by equally well-designed monitoring and control technologies.

The number of remotely operated aircraft is increasing. The Pentagon and Central Intelligence Agency (CIA) admit to deploying 19,000 drones (2014), and many more are used in other countries. The United Kingdom, Pakistan, China and Israel have joined the United States as war ready with large stockpiles of drones that can deploy weapons. Australia is planning to join this cartel. Many more countries have huge numbers of drones performing tasks as varied as exploring remote areas by camera and monitoring bushfires and traffic flows. They are used for filming by movie makers and TV companies. Small drones make great toys for children.

Pilotless drones have been around since the First World War, but their use is increasing dramatically as governments, law enforcement agencies and private companies find new uses for them. The challenge for the future will be to design and develop systems for tracking their movement and ensuring they do not collide with other drones or piloted aircraft.

Mines are installing driverless trucks and coal cutters. In the quest for greater profits, mining companies are investing in automated machinery that can be operated 24/7 and reduce their spending on expensive operators. Observers of this phenomenon predict the day will soon come when large mines will operate with automated equipment and just a few maintenance staff.

MEASURING COMPLEXITY

In order to assess the risk potential of a design, we need to measure the overall level of complexity. Through careful analysis, it should be possible to arrive at a figure that indicates a grading of potential risk of a new or reworked design. This analysis should also be able to indicate individual features that can be modified to lower the overall risk tally.

Researchers have devised equations to measure the degree of complexity in ICT systems. Much of this is based on the number of interactions within a section of code. It is believed that the more interactions, the more chance of design error through faults in the algorithmic basis underlying the code. There is also the opportunity for faulty data to corrupt the interactivity within the code.

What is not currently being assessed is the added complexity when humans interact with the system. Work is being done on a design process analysis tool to identify and measure error potential in complex systems taking into account not only the physical design, but also the human factors issues when humans interact with the design product. See the proposal at the end of Chapter 11.

AND THE PROBLEM IS?

What is the problem with increased automation? More complex systems have more variables. The more variables, the harder it is for them to be traced individually. The harder it is to trace the operation of individual variables, the harder it is to identify the interactions with other aspects of the system. The harder it is to identify interactions, the more difficult it is to remedy anomalies and ensure a smooth and error-free operation of the system. Let us take the case of a computer system to control the movement of trains.

In its simplest form, the software needs to be able to identify where the train is, operate signals at the beginning and end of each block, move points to allow the train to cross from one track to another and warn the driver of track repair work or other conditions that require a change of speed. It may also need to record and compare the axle counts from the beginning and end of the block to ensure the whole train is still together. Such a device is used with heavy coal trains in Australia. We already have a bevy of variables.

Let us now consider the removal of the driver. The system must now be able to 'see' the track and stations. Video recordings need to be analysed and converted into computable data that can add to the system's moment-by-moment knowledge of the train's progress, track condition, movement of passengers at stations and so on. Decisions about speed and safety must now be made by the system instead of the driver.

Consider the movement of 100 driverless trains on the same network. The system now has to know where each train is and operate signals, points and interlocking, and control the speed of each train to prevent collisions. For instance, if one train has a problem and stops, other trains on the same route must be diverted from that track. Consider the overlay of visual data on top of normal running data. The variables have increased exponentially. What will happen when someone who is running late to catch the train gets one foot inside the door when the doors close automatically?

Norman (1988) argues that the problem with automation is that it is inappropriately designed. He holds that the level of intelligence built into an automated system is powerful enough to take over normal operation, but is not enough to cope with abnormal situations. He believes that automated systems should be made either more intelligent, so they can cope with every possible situation, or less so, to allow operators to deal with inconsistencies.

As the situation becomes more complex, the chances of design error increase dramatically. When an error occurs, it becomes harder to trace the first cause, and the accident is likely to be more spectacular. Most often, human error is blamed, but in many cases the first cause is a problem in the design process.

We know that automated systems, no matter how complex, are not as intelligent as humans. As long as they operate within a clear set of guidelines, they can cope very well – in fact, in some cases better than humans – but when something unusual happens, they have no ability to deviate from their programmed behaviour. They cannot assess an unusual situation and vary their responses to cope. If systems are to replace people, and this is happening in many spheres, they need to be made more intelligent. They need to be programmed to deal with unusual situations, and to do this, designers need to spend time with end users, gathering knowledge of likely unexpected events, and building in strategies to assist complex systems in coping in unusual situations. Until that day arrives, automation should not replace humans entirely. Machines can deal with drudgery and repetitive tasks extremely well, and operate without being tasked by the boredom these activities produce in humans, but until they can deal with unusual stresses with the same behaviours humans can call on at a moment's notice, they should be recognised as useful companions, but not overlords.

THREE REASONS FOR THE ADVANCE OF AUTOMATION

Three factors stand out when considering the relentless movement of automation into almost all aspects of our lives. The first of these is the irrepressible and addictive human need to change the way we do things, improve tools and speed up processes. Early humans sharpened sticks to make spears. Then someone found that pieces of flint could be fixed to the point to make a sharper entry into an animal's body. Someone else began adding barbs near the point to help the spear stay attached to the animal. Others found that the spear could be thrown farther with a throwing stick, or woomera, or a shortened spear could be sent even farther by using a bent stick and a length of vine. Inventors everywhere are continually designing new ways to carry out tasks.

The second factor is the mistaken belief that when things go wrong, it can almost always be attributed to human error, most often on the part of the end user. I studied 76 accident databases from around the world, including air and rail crashes as well as accidents involving land and water vehicles, and industrial disasters too. The majority of accident reports begin by blaming the end users, the pilots and engineers, the captains and drivers, the managers in charge of industrial sites or the workers at the 'workface'. But further investigation revealed that many accident investigators admitted the first cause of many of these accidents was likely to have been a design error. Something went wrong in the design process of ICT systems and technologies, engineering designs or the design of work procedures and processes. The media is as guilty as companies of promulgating the myth of user error – the blame game, the scapegoating.

The third factor is financial. Human workers are expensive. They need wages, paid sick leave, public holidays and vacations. Their work output varies according to their health, their emotional state and their attitudes towards the job, managers and their fellow workers. It makes financial sense to replace them with robots or other technologies where the task is a repetitive one. Robots can work 24 hours a day, 7 days a week, 52 weeks a year. Their installation cost is high, but ongoing maintenance is low. Once operational, they are much cheaper and more productive than humans doing the same task. The huge limitation of tooling up a robotic process is the time and money required when a change to production routine needs to be made quickly to meet changes in market demand.

TWO SOCIAL IMPACTS OF AUTOMATION

The first social impact of automation is the dislocation of the workforce. This dislocation can be seen as a slowly advancing phenomenon that began many years ago. One of the reasons the United States and then Australia were used as convict dumping grounds was the enacting of the Enclosure Acts by the British government between 1604 and 1914 (www.parliament.uk). Land owners were permitted to fence common land and open fields in villages, removing the ability of peasants to gain a living in country areas. They fled to the cities in thousands and existed any way they could. This often meant stealing a loaf of bread or an apple or two, for which they ended up in prison. Many were eventually transported to the United States and later to Australia.

Many other trades have disappeared over the years. Buggy makers lost their livelihood when cars were introduced. Can you remember when garbage trucks had a team of council workers who ran behind the vehicle, lifting garbage bins from the pavement and emptying them into the back of the truck? Now one driver uses a mechanical lifter to empty each bin.

I interviewed a rail system controller who recounted that at a busy railway station near his home, there used to be seven attendants, one at each exit gate to collect tickets from those alighting from the train. They were replaced by a number of machines that allowed you to pass once you fed your ticket into the slot. His issues were that, not only had seven people lost their jobs, but the machines were placed across one exit and the other six exits were closed. He was concerned about the crowding in the evenings when hundreds of workers were forced to queue for the remaining exit. During the evening crush, he had seen people pushed over the yellow safety line towards the track. He believed it was very likely that someone could be pushed to their death in front of an oncoming train.

So we have workers, many of them highly skilled, losing their jobs. If they are lucky, they may be able to retrain and work in a different field. If they are unlucky, they will join the dole queue. This kind of dislocation is emotionally draining and can have a huge impact on that person's family. Most still have mortgages to pay and may be forced to sell their home and find somewhere cheaper to live. The social impact for these people is huge.

The second major social impact of automation is the intensity and magnitude of accidents when automated systems go wrong.

Airbus A320 Crash

This crash at the Habsheim air show in France in 1988 offers us a classic illustration of what can go wrong when the pilot is designed out of the cockpit. This first fully automated aircraft was conducting a low-speed flyover with landing gear down. This manoeuvre should have been conducted 33 metres off the ground, but the plane was flying at around 9 metres. It skimmed the tree tops of a nearby forest and crashed into the ground, killing three passengers.

Driverless Monorail Train Crash

At least 23 people were killed and 10 injured when a driverless monorail train in Germany ploughed into a maintenance truck at 200 kilometres per hour (Harding 2006). Most of the passengers were in the front carriage peering out of the panoramic front window. They saw the impending crash but could do nothing to slow the train. A company spokesman said he believed the accident was due to a communication breakdown. One has to ask why the train was not equipped with forward vision cameras that could detect an obstacle on the track and activate the braking system.

Rail Crash in Spain

In Spain in 2013, an automated train travelled around a bend at 190 kilometres per hour when it should have slowed to 80 kilometres per hour. The derailment and impact with pylons holding up a road bridge over the track killed 80 people and injured 130. The attendant in the cabin was blamed and initially accused of murder, but it was eventually admitted that the automatic control system did not apply the brakes at the appropriate time (Abend 2013). An accident investigator commented, 'Either the wrong system was used or there was a flaw in the system'.

We see an increasing list of serious accidents, very often caused by design error. It is understandable that increasingly complex technologies mean a greater number of variables. Variables that are not recognised and dealt with safely in the design could potentially cause safety hazards. In order to understand design error potential and take steps to make designs safer, we need to identify how and where errors can occur in the design process, and take steps to mitigate those errors.

FUTURE DIRECTIONS IN AUTOMATION

The innovation with potentially the greatest impact on our lives in the future is the automated car, the self-drive vehicle that will take care of our commuting, allowing us to work, play or talk with our family and friends or business contacts as we are transported by the machine from home to office, school or shopping centre.

Driverless cars are already a reality. A fleet of self-driving vehicles already ferry passengers from Heathrow's business park to Terminal 5. Volvo already has 5 automated cars on the roads of Gothenburg and is planning to have 100 within the next two years. The Volvo company is also conducting tests of their automated vehicles in Adelaide, South Australia. The government of that state is hoping to replace their closing traditional car manufacturing centres by leading the country in the production and distribution of automated cars.

Google has built its own prototype driverless cars and is looking to go into production with a local automotive company. Press releases suggest they are talking with Ford. Mercedes Benz has unveiled its first driverless car at the consumer electronics show in Las Vegas.

These vehicles at the moment travel slowly, perhaps at only 40–80 kilometres per hour. Some of them still have a steering wheel, an accelerator and brakes, so control of the vehicle can be taken over by a human in an emergency. Some of them are completely automated and have no steering wheel, accelerator or brakes. Once the passenger has indicated a destination, the vehicle takes care of the rest, or in the case of the Heathrow vehicles, the destination has already been programmed in. It is held by the designers that all-round sensors will ensure these vehicles never collide.

CHANGING LIFESTYLES

The reason that this is such a big deal is because a car that can drive itself is going to change how and where we live and work. The vehicle can dynamically position itself to the point of need and then do work by moving people or goods to where we instruct. We will eventually have a limitless supply of on-demand chauffeurs, or fully automated taxis, if you prefer.

We will not really need to own a vehicle anymore if we do not want to, or perhaps own several, as the automaker's business model would prefer. Rather than have our multipurpose vehicle sit idle some 90% of the day, we can hire a vehicle suitable for our needs, a shared self-driving electric two-seater for our commute (and work on the way) and an SUV for the group trip at the weekend.

Some people will prefer to live farther from the city, as the commute can now become part of their working day, allowing them to work while travelling. Some will find it even easier to live in the urban centre where they need never own a car again. Small businesses may find it cheaper to use their self-driving vehicle as their office rather than rent space in a building. As the car drives to their next appointment, they can phone clients or type, print and review contracts and other business documentation.

Some will find their jobs start to change or even disappear. There will be enormous changes for people such as taxi, truck and bus drivers. And life will change dramatically for auto body repair staff, auto insurers, road safety professionals and transportation planners.

New jobs will appear as new business models spring up overnight. This novel technology will be the catalyst for many new societal interactions and services. The hospitality industry will be pleased if it means more people are willing to get out for the evening and have a good time because they are less concerned about how they will get home. Will it be possible for people who drink over the alcohol limit to be fined when they are merely a passenger in an automated car and not the driver?

Self-driving vehicles will greatly improve access to business premises. People who are disabled, too poor, without a driving license or unable to drive will now find it much easier to hold down a job.

INTERNET OF THINGS

The Internet of Things (IoT) is a recent title for what used to be called machine to machine (M2M). The 'things' are, in the main, sensors that communicate with computers and computer-driven technologies, passing on information for those systems to take appropriate action. A simple example, and one in almost every home and office, is a temperature recorder, a sensor that measures and relays the room temperature to the air conditioning apparatus, which will turn itself on or off or moderate its settings when the temperature rises or falls outside preset levels. The use of sensors connected to 'intelligent' machinery is expanding at a phenomenal rate with the advance of a form of automation that is all about things operating automatically without the intervention of humans, but let us go back to the beginning.

AN EARLY M2M SYSTEM

Some years ago, one of my sons was hired by a large company that owned a string of around two hundred 5- and 10-cent stores in Australia. He was engaged to head the linking of those stores electronically with the central warehouse.

At that time, at the end of each day's trading, a staff member at each of those stores would fax a list of every item that had been purchased from that store that day to the warehouse. The next day, warehouse staff would work from those faxes to gather replacement items for the next delivery to each store.

His task required replacing the old cash tills with new electronic registers and linking them with a large computer in the warehouse. Each sale could then be relayed electronically from the register to the central computer, which compiled and printed out stock replacement lists for each store.

Communication between each cash register and the central computer took place continually during opening hours, removing the tedious handwriting of fax lists and overloading the warehouse fax facilities after store closing time. This early M2M communication methodology provided a more efficient data transfer strategy, saving tedious hours of stocktaking and faxing at the stores, and similar tedious hours of fax interpretation and verification when things went wrong (such as a fax machine running out of paper).

Since that time, M2M has taken huge strides. In many places, wireless communication has replaced wired, and the data transfer from register, ATM or other vending machine has been linked to warehouse management software, remote control, robotics, traffic control, logistic services and supply chain management. Automated functionality has been applied to conveyor belts, belt scanners, wrapping and packaging machines, palletting and other goods handling and sales preparation tasks. These linked systems have brought automation to warehouse operations. This automation has removed many of the tedious and time-wasting manual tasks, greatly reducing the number of staff required to pull orders by hand, package, label and dispatch.

The word 'complexity' warns us there will be many variables operating at any one time, and the large number of variables opens the door to design errors. We have all seen those cartoons where an automated process gets out of control and cannot be stopped. Imagine a delivery belt in a warehouse spewing out carton after carton of an item, quickly building a small mountain of boxes. Impossible? Not if a glitch appears in one of the systems. Unfortunately, the cartoon joke can and does turn into disastrous reality.

Businesses are beginning to make use of third-party applications in the cloud rather than depend on their own local ICT department to produce the computer systems necessary to drive M2M. It is believed these systems have been more thoroughly tested and proved by other uses around the globe. This does not completely negate the problems of unsafe design because local modifications are necessary to configure the cloud systems for the local operation.

In recent years, M2M has moved from warehouse operations to other areas. We can find M2M working within our modern cars where inbuilt sensors send messages about engine operation problems, or relay information about engine and electric motor operation in hybrid vehicles. Medical devices can measure critical functions like blood pressure and send that information directly to the doctor. Sensors can be found in trains and aircraft, dam walls and manufacturing machinery, shipment refrigerators and containers. They can measure temperature, weight, vibration or noise levels, to name just a few. The data from these sensors can be sent to offices and homes, engineering and transport centres or control rooms. Other computers receive the data and respond appropriately if the readings are outside the allowable limits. Sensors can also monitor the health of other machines and report operational problems so techs can fix them.

One of the M2M systems I have worked with is the supervisory control and data acquisition (SCADA) control software. In control rooms, operators use the SCADA system to control switches and other devices out in the field. Power company controllers can read voltage flows and other information communicated to them by sensors and, when necessary, re-route high-voltage electricity flows from one feeder to another by remotely operating switches in sub-stations. When preset limits are breached by storm or accident, alarms will sound in their SCADA system to alert them to a problem that needs their attention.

M2M has brought efficiencies and bottom-line benefits to the whole supply chain for many companies. But, it has also brought some issues that need answers. The first is the replacement of human jobs with technological systems. Large numbers of workers have been dispossessed. Another important issue with this rapid growth in M2M, particularly in regard to cloud technology, is security. How can the data be kept private, protected from infiltration and safe from integrity loss?

TARGET WAS TARGETED

Foster (2014) reports three serious security breaches involving M2M systems. The first happened when hackers installed malicious software in the checkout system of Target's more than 1800 U.S. stores. This resulted in some 40 million credit and debit cards being compromised and the loss of personal information relating to as many as 70 million customers. The entry was apparently made through the M2M intelligent, interconnected building management systems. From there, the hackers were able to move to the company's point-of-sale (POS) systems. The breach appears to be the result of the company not properly segmenting its data networks.

PROBLEMS WITH MEDICAL DEVICES

The second case relates to the hard coding of passwords into more than 300 medical devices. The U.S. Food and Drug Administration issued an alert, urging medical device manufacturers to upgrade security protections. It was reported that former U.S. vice president Dick Cheney's doctor insisted the manufacturers of his pacemaker disable its wireless capabilities so that an assassination attempt by electronic signal could be prevented.

STUXNET

In 2010, the Stuxnet virus infiltrated Iran's nuclear facilities in Natanz, destroying 1000 fuel refining centrifuges. Stuxnet was aimed at programmable logic controllers (PLCs) that are used in many supervisory control and data acquisition systems around the world. This illustrates the vulnerability of M2M to malicious attack.

M2M, under its new name of the Internet of Things (IoT), is bringing automated devices to us at a rapid rate. So many tedious tasks will be handled by machines. But, there are dangers in this. The first is that many people are losing their jobs as machines take over tasks formerly carried out by humans.

The second is that we are being de-skilled, and when these systems fail, as they do, we will eventually be unable to repair them quickly or devise workarounds to keep them operating.

The third, and potentially most worrying, aspect is the increasing complexity. We know from sad experience that increased complexity means increased variables, and increased variables means greater potential, not only to fail, but also to cause accidents that can hurt or kill many people. Designers will need to work very hard to build safety into new technologies, and to do that, they have to look very carefully at the design processes they use and modify their approaches. Future design will depend much more on using participative design models where stakeholders and end users become part of the design team and become heavily involved in design concept formation, testing and implementation.

It is now clear that hackers can enter IoT systems in businesses and homes and create mayhem. It has been demonstrated that hackers could electronically enter some advanced cars and take control over brakes, steering and engine operation. Hewlett-Packard released a study that concluded that 70% of IoT devices are vulnerable to hacker attack. Designers need to take heed of these warnings and work harder to produce safe designs.

**

The development and rapid expansion of M2M/IoT technology is beginning to bridge the gap between automation and artificial intelligence (AI). A discussion of AI, perhaps the most complex of all forms of automation, and certainly the one with the most far-reaching potential to dispossess humans, can be found in Chapter 10.

10 How Artificial Is Artificial Intelligence (AI)?

> The development of full artificial intelligence could spell the end of the human race.
>
> **Stephen Hawking**

More than a hundred years ago, E. M. Forster (1909) wrote a story called 'The Machine Stops'. He imagined a world that was controlled by a single computer system known as 'the Machine'. Slowly, technicians were phased out as the Machine took care of its own maintenance. Eventually, sub-systems began to fail because there were no longer any technicians who knew how to repair them. People learnt to lower their expectations. The Machine had taken the place of God, so there was no criticism.

When the Machine stopped, the world's population perished because people had lost the ability to care for themselves. They had become completely deskilled. It is not hard to imagine the current world spiralling into a planet controlled by advanced technology to the detriment of the populace.

Despite Forster's prediction and Hawking's concern, efforts to develop artificial intelligence (AI) continue, with researchers devoting time, energy and money to making computers behave like humans. Some of them are succeeding: machines can now understand humans, speak with them, learn from them and write like them. They will make some jobs obsolete and others easier. But we are not in danger – just yet. At the moment, AI is a joint traveller with other automated technologies, but it will eventually outstrip and control them all.

HAS AI ALWAYS EXISTED?

From earliest times, there was a fascination with humanlike images that were believed to have intelligence. Thinking machines and artificial beings existed in ancient Egyptian and Greek mythologies. Cult images, some of them animated, were built in every major civilisation. By the nineteenth and twentieth centuries, artificial beings had become a common feature in fiction, including Mary Shelley's *Frankens tein* and Karel Čapek's *R.U.R.* (*Rossum's Universal Robots*). Some writers attribute this to a desire to create 'gods' or God-like images.

In the 1970s, the first attempts to develop electronic AI systems were thwarted by lack of sufficient computer memory. Asking a computer to act like a thinking being requires an enormous amount of processing power, and those early computers just did not have the 'grunt'. Along with others, I experimented at that time with what were called 'knowledge-based' or 'rule-based' systems. They later took the name 'expert systems'. As computer power increased, AI became possible.

In the 1990s and early twenty-first century, AI achieved its greatest successes to date. AI was shown to have applications in logistics, data mining, medical diagnosis and many other areas throughout the technology industry. The success was due to several factors: the increasing computational power of computers, a greater emphasis on solving specific sub-problems and the creation of new ties between AI and other fields working on similar problems.

In 1997, *Deep Blue* became the first computerised chess-playing system to beat a reigning world chess champion, Garry Kasparov. In 2005, a Stanford robot won the DARPA Grand Challenge by driving autonomously for 131 miles along an unrehearsed desert trail. Two years later, a team from Carnegie Mellon University won the DARPA Urban Challenge when their vehicle autonomously navigated 55 miles in an urban environment while adhering to road rules and avoiding traffic hazards. In February 2011, in a *Jeopardy!* show exhibition match, IBM's question answering system, *Watson*, defeated the two greatest Jeopardy champions, Brad Rutter and Ken Jennings, by a significant margin. The Kinect, which provides a three-dimensional body–motion interface for the Xbox 360 and Xbox One, uses algorithms that emerged from lengthy AI research.

ALAN TURING, THE FATHER OF MODERN-DAY MACHINE INTELLIGENCE

Mechanical or formal reasoning has been developed by philosophers and mathematicians since antiquity. This led to the invention of the programmable digital electronic computer, based on the work of mathematician Alan Turing and others. Turing suggested that by shuffling the symbols 0 and 1, a machine could simulate mathematical deduction. The 0s and 1s were created by simple switches where 0 meant the switch was off and 1 meant the switch was on. These switches (called bits) were arranged in groups of eight (called bytes), and the way they work together is governed by the 'binary' code. In this code, 00000001 = 1, 00000010 = 2 and 00000011 = 3. If all the switches were on, for example 11111111, the digit value would be 255. Each step to the left is the next power of 2 (e.g. 128, 64, 32, 16, 8, 4, 2, 1, which added together make 255).

A method was devised for simulating language using these zeros and ones. For instance, the lowercase 'a' can be represented by '01100001' in binary code, which is equivalent to '97' in our normal decimal notation. Thus was built a code where a string of eight zeros and ones could represent letters of the alphabet as well as digits and other text signs, such as @, $ and ?. This American Standard Code for Information Interchange (ASCII), as it came to be known, was then used to create simple computer programs.

Turing formulated a test by which a machine's responses could be called 'intelligent' or not. In this test, a man (A) and a woman (B) go into separate rooms and guests try to tell them apart by typing a series of questions and reading the typewritten answers sent back. In this game, both the man and the woman aim to convince the guests that they are the other. Turing then asked, 'What will happen when a machine takes the part of A in this game?' Will the interrogator decide wrongly as often when the game is played like this as he or she does when the game is played between a man and a woman?

Later, Turing suggested an alternative formulation of the game which involved a judge conversing only with a computer and a man. If the judge cannot reliably tell the machine from the human from typewritten responses to questions, the machine is said to have passed the test. In another version, a jury asks questions of a computer and the role of the computer is to make a significant proportion of the jury believe that it really is a man.

Researchers in the United Kingdom had been exploring 'machine intelligence' for up to 10 years prior to the founding of the field of AI research at the Dartmouth Conference in 1956. The conference organisation was spearheaded by Marvin Minsky and attended by some of the people who would become leaders in AI research.

KNOWLEDGE-BASED SYSTEMS

These early attempts at producing AI systems had two components, the knowledge base, which was the database holding the responses from a respondent, and the inference engine, which was a set of rules used to analyse each response and convert it into the next question.

Lack of memory in earlier computers precluded any intelligible language interpretation, and we did not have voice recognition, so everything had to be displayed on the screen in the form of text. One of the first systems was able to emulate a Rogerian style of psychological analysis where the computer would display a message like, 'What concerns you?' A typical 'conversation' might look like this:

Computer: What concerns you?
Respondent: My mother doesn't like me.
Computer: Why doesn't your mother like you?
Respondent: She prepares food I don't like.
Computer: Why does she prepare food you don't like?
Respondent: She says it is good for me.
Computer: Why is it good for you?
Respondent: I don't know. She just says it because she doesn't like me.

EXPERT SYSTEMS

Over time, as computer storage and processing power grew, these systems could store information from subject matter experts (SMEs) to populate the database. From then on, they were known as 'expert' systems. One widely acclaimed expert system held information about the diagnosis of tropical diseases and was used by GPs not familiar with tropical medical problems to help diagnose complaints brought to them by patients who had visited a tropical location. The system not only helped identify the disease, but also suggested appropriate medication, going as far as computing the dosage based on gender, age and weight of the patient. At this time, ICT experts developed what was termed 'fuzzy logic'. As an example, the tropical disease system could take the responses from the GP and calculate that there was, say, an 80%

chance that the patient was suffering from X disease and a 65% chance that it was Y disease. Further tests or amended treatment would then be suggested.

MYCIN

MYCIN was an early expert system that used AI to identify bacteria causing severe infections, such as bacteraemia and meningitis, and to recommend antibiotics, with the dosage adjusted for the patient's body weight. The name was derived from the antibiotics themselves, as many antibiotics have the suffix '-mycin'.

MYCIN was developed over five or six years in the early 1970s at Stanford University. This system was never actually used in practice, but research indicated that it proposed an acceptable therapy in about 69% of cases, which was better than the performance of infectious disease experts who were judged using the same criteria.

MYCIN operated using a fairly simple inference engine that judged the data using a knowledge base of approximately 600 rules. It would query the physician running the program via a series of textual questions requiring only yes/no responses. At the end, it provided a list of possible culprit bacteria ranked from high to low based on the probability of each diagnosis. The reasoning behind each diagnosis and a recommended course of drug treatment was given.

ELIZA AND PARRY

In 1966, Joseph Weizenbaum created a program which appeared to pass the Turing test. The program, known as ELIZA, worked by examining a user's typed comments for keywords. If a keyword was found, a rule that transformed the user's comments was applied, and the resulting sentence was returned. If a keyword was not found, ELIZA responded either with a generic riposte or by repeating one of the earlier comments. Weizenbaum developed ELIZA to replicate the behaviour of a Rogerian psychotherapist using a technique that kept asking questions based on earlier replies.

The program was able to fool some people into believing that they were talking to a real person, thus supporting the claim that ELIZA was perhaps the first program able to pass the Turing test.

Kenneth Colby created PARRY in 1972, a program described as 'ELIZA with attitude'. It attempted to model the behaviour of a paranoid schizophrenic, using a more advanced approach than that employed by Weizenbaum. To validate the work, PARRY was tested in the early 1970s using a variation of the Turing test. A group of experienced psychiatrists analysed a combination of real patients and computers running PARRY through teleprinters. Another group of 33 psychiatrists was shown transcripts of the conversations. The two groups were then asked to identify which of the 'patients' were human and which were computer programs. The psychiatrists were able to make the correct identification only 48% of the time – a figure consistent with random guessing.

It was reported that a psychologist in the United States ran a program of this type in his waiting room. By the time his patients saw him, the specialist reported, they had largely sorted out their own problems. I am sure he still charged his hefty fee.

ARTIFICIAL INTELLIGENCE TODAY

Today, AI can do so much more because of the vastly increased speed and storage space in modern computers. It is used to scan our cheques at the local ATM and help us reach our destination through our car global positioning system (GPS). Robots have become common in many industries. They are often given jobs that are considered dangerous to humans. Robots have proven effective in jobs that are repetitive, which may lead to mistakes or accidents due to a lapse in concentration by humans. AI drives the avatars that can be seen on Web pages and in computer games.

AI methods are used in pattern matching, for example recognition of optical characters, handwriting, speech and faces. It is also used in computer vision, medical diagnosis and natural language processing, and is important in robotics, automation, training simulators and concept and data mining. Many of the processes we now see in the Internet of Things are controlled by AI systems.

NATURAL LANGUAGE PROCESSING

The first mainstream consumer application of a machine that can comfortably interact with humans in something close to natural language came from Apple in the form of the *Siri* personal assistant. Since then, Microsoft has launched *Cortana* and Google also gave users the option to talk to their phones.

But, Siri and its peers are not really AI. Though natural language has made great leaps in the past half-decade, it remains limited.

For example, you can ask Siri to find restaurants in a particular neighbourhood. But, Siri cannot engage in more complex interactions that involve multiple data points, such as 'I'd like a reservation for a restaurant in Brisbane, preferably Italian, but Thai will do, for five people at around 7:00 p.m. Oh, and I'd like one with on-site parking and a children's menu'. That kind of request requires multiple interactions that Siri and its partners are not yet ready to perform.

AMELIA GOES TO WORK

While a machine that humans can casually chat with may not be here yet, a more basic, text-based version for business is already available. IPsoft, a New York–based technology firm, has developed what it calls a 'cognitive knowledge worker' named *Amelia*.

Amelia's anthropomorphic avatar is a pleasant, blonde woman with neat hair and a somewhat intense gaze. She has the comprehension of a six-year-old and an emotional range to allow for appropriate reactions to the tone of her human interlocutor. If a human is becoming irritable, Amelia adjusts her tone accordingly. And if she cannot solve a problem, she calls in a human operator – and then watches and learns from that interaction.

When a human types a message to Amelia, the software breaks the message down into its component parts. As the conversation progresses, Amelia is able to relate prior questions to the current ones, an impressive feat for a machine.

YSEOP

Another company, called Yseop, uses AI to turn structured data into intelligent comments and recommendations in natural language. Yseop is able to write financial reports, executive summaries, personalised sales or marketing documents and more at a speed of thousands of pages per second and in multiple languages, including English, Spanish, French and German.

ARTIFICIAL INTELLIGENCE IN THE FUTURE

Views on where AI is going are mixed, but mostly positive. Some of the responses to this question were

- 'Robots will keep us safer, especially from disasters'.
- 'AI will allow us to address the challenges in taking care of an ageing population and allow much longer independence'.
- 'It'll enable drastically reduced, maybe even zero, traffic accidents and deaths'.
- 'AI will become an assistant for humans, making what humans want to do and what humans want to be easier to achieve with help from AI. What if I lost a limb and I can't swim as fast? What if an AI can actually know how to control this robotic limb that's now attached to me?'
- 'Future systems may work via augmented reality. For example, I hope that exoskeletons will allow me to walk when I am old and feeble. I hope that I can retain my sense of hearing and sight even as my eyes and ears fail'.
- 'Very smart computers could solve all our problems, including climate change'.
- 'AI is embedded in many of the technologies that have been changing our world over the last several decades and will continue to do so'.
- 'As machines get more intelligent and can better adapt to their "users", people may end up preferring to deal with machines than with people. They might even become our best buddies'.
- 'AI could also give us more time to be creative'.

On the other hand, some people see negative outcomes, such as

- 'Becoming cyborgs. Imagine if we could augment our brains with infallible memories and infallible calculators. What's going to happen when we have this kind of AI but only the rich can afford to become cyborgs? What's that going to do to society?'
- 'I think we'll see profound changes in the nature of work. It's hard to think of a job that a computer ultimately won't be able to do as well if not better

than we can. This will force profound changes within society. Will we work a shorter working week? Will we work at all? How will we distribute the wealth that this generates? This is a challenge not for scientists but for society'.

WHY IS AI DANGEROUS?

Much is written about the question of whether AI can be developed in such a way that the intelligence of robots and other machines of the future will reflect an approximation of the emotive side of human nature. Should machines be taught to believe in right and wrong? Can they be developed in such a way that they act from a moral set of values? If this can be done, will they judge human behaviour and act to discipline or remove humans they see acting contrary to the set of rules in their inference engine?

We could argue for hours about whether the development of self-awareness necessarily means the development of human qualities like egoism, compassion and greed. And this is the area where Steven Hawking and others express concern. Our fear of AI seems to stem from thinking about the advances in AI as a projection of ourselves. But, robots are not like us. They do not have a life span and so have no sense of time. They do not need to eat or drink, live in a home or take holidays, so money and other material goods have no meaning. Possibly the only emotion they could share with us is when there is a threat to their existence.

My personal concerns about the development of AI relate to the way their mechanical and electronic systems, together with their databases and inference engines, have been constructed. We should be aware of what information and values are planted in their memory areas, and what design errors have been allowed to enter their components. We know that complex systems have an enormous number of variables. We also know that greatly increased numbers of variables and the myriad interactions between them open the way for design and operational error. Those who design these systems need to be extremely vigilant and check and recheck for potential faults at every stage of the design process.

**

The rules outlined in Chapter 11 should be just a starting point for the most intense scrutiny when designing the AI robots, machines and systems of the future. One mistake could have unbelievably serious consequences.

11 The Solution Is …?

There is no simple solution to mitigating design error. Systems, devices and processes are becoming more complex and the number of variables increasing at an exponential rate. Most of the rules for safe design are quite similar for all design, but we do need to make some adjustments depending on whether the design is guided by a RED CAP, BLUE CAP or YELLOW CAP systems development life cycle (SDLC) model.

RED CAP STRATEGIES

We discussed earlier the difference between huge design projects such as major bridges, high-rise buildings and space flight vehicles and trajectories, to name just a few of those projects that require enormous resources and planning, and the smaller projects that are built much more quickly.

The big projects, the RED CAP group, are those where the end result has to be set in concrete before work begins on the design. The broad-brush outline might look a little like this. We need to build a multistorey apartment block with a base of 55 by 80 metres and a height of no more than 220 metres. It is to have lifts to each floor at each end of the longer side and safety stairways circling the lifts. What often happens is that stakeholders and end users too are not involved, except perhaps to view a model of the complete design. At this stage, it is often too late for them to indicate special features that should be included, safety issues and information concerning the locale and the expected client group that should be taken into consideration.

It is essential that RED CAP projects include stakeholders and end users, or at least knowledgeable user experience (UX) professionals, during the concept formation stage. It is also expected as a matter of course that the concept design team will include subject matter experts (SMEs) who have specific knowledge related to this kind of project. I would also expect human factors people and safety officers who can bring to the table their particular expertise. I cannot imagine designing a rocket to take people into space without including astronauts who can offer advice on the layout of cabins, the movement pathways between areas of the cabin and specialised equipment placement.

Once the concept has been agreed, recorded in specifications and the whole design team has signed off on the proposals, I expect that the designers, developers and engineers would go ahead in building the design according to the specifications. I recommend that each design team have independent auditors who can view progress and ensure the agreed concept is being built.

For this kind of project, one of the traditional SDLC models, such as *waterfall*, will be used and followed to the letter once the design concept has been signed off. Each step needs to be carefully tested and a thorough whole system test carried out before handover.

Rules for a Successful RED CAP Design Project

The rules required for this type of design include

1. Gather full information on the project before beginning. This will mean consulting with the funding body, perhaps government or local council representatives, unions, safety experts and SMEs to make sure the necessary rules and regulations will be complied with. Information should also be gathered on previous similar projects, together with a report on any problems, deficiencies or errors experienced with those earlier designs.
2. Gather a design team that includes stakeholders, end users, UX and safety experts, SMEs and any others who can contribute to the design concept. Make sure you cover every aspect of the design, including its suitability for the location, appeal and nuisance value for neighbours.
3. Make sure the whole team has the opportunity to contribute equally to the discussions and decisions about aspects of the design.
4. Record every detail of the design in complete and full specifications.
5. Have the whole design concept team sign off on those specifications before work begins. At this stage, the stakeholders and end users have done their job and may be dismissed.
6. Employ an independent audit group to oversee the building of the project, checking against the specifications at every step.
7. Carry out a continuous unit testing schedule during the build and a complete end-to-end test of the design before delivery.

BLUE CAP STRATEGIES

You may remember from earlier discussion that BLUE CAP projects are prototypes. Someone has an idea and needs to test whether it has any value. I remember a project I was involved with some years ago where the design team had decided to modify an existing stand-alone computer system so that it could be accessed via the Web. None of the developers in this team had ever worked on a Web project before, so a Web programmer was called in to build a prototype of the front end of the existing system to show what it might look like as a Web page. The Web designer spent two weeks building his model. The team liked what they saw and decided to go ahead with the project. The prototype was there to view, but it was not modified and not included in the final project.

Rules for a Successful BLUE CAP Design Project

1. Prototypes are used to test design ideas. They are not meant to be any more than an example of what might be achieved.
2. Once a prototype is accepted, it is time to change to a YELLOW CAP strategy to plan and build the final design.

YELLOW CAP STRATEGIES

YELLOW CAP strategies suit most design projects. They are normally short-term projects that need to be planned, built, tested and deployed as quickly as possible. The projects suitable to YELLOW CAP strategies include most information and communication technology (ICT) systems, engineering projects and the creating of business processes and procedures.

Rules for a Successful YELLOW CAP Design Project

The following rules capture the lessons learnt from recent research that can help to produce safer designs:

1. Use an *adaptive* SDLC model, or if you need to use a traditional model, make sure it is flexible enough to allow stakeholder and end-user input at every stage.
2. Insist on the participation of end users in the design process – in fact, in every step in the project.

Concept Formation

If the design concept is flawed, it is likely that the design stages that follow will be flawed too. A design is normally called for to solve a problem, so it is important for designers, and the managers who enlist their assistance, to be quite sure of the parameters of the problem. They need to listen to end users because they know the operational needs and can provide valuable insight into solving the problem.

During the concept discussions, allow for brainstorming, mind mapping and similar strategies so that all possible problems and solutions are aired. Make sure every person contributes. Do not allow the highest paid person's opinion (HIPPO) to sway decisions.

3. Fully investigate the problem.
4. Investigate each possible solution thoroughly.
5. When a design concept is agreed, have all members of the team sign off before proceeding any further.

Specifications Writing

Specifications documents translate the design concept into clearly articulated instructions to guide the development of the system, device or procedure. For best results, skilled technical writers should be employed so there are no gaps. Often, there will be more than one specifications document. Some of the adaptive SDLCs minimise the creation of binding specifications at the beginning of the project. This means that technical communicators need to record every discussion and interim decision as development proceeds so that a clear and detailed record of those discussions and decisions is available at any time during the build.

The list often includes system, technical and user specifications, and may also include a data dictionary so everyone on the team understands the precise meaning of terms used. Many systems and procedures call for reports, so make sure the structure and content of reports is accurately described.

There may also be a style guide or template that sets out the way the final product should be presented. This normally includes company branding, logos and use of colour, as well as screen or paper layout, number of columns, typeface and font.

6. Testing is done against these specifications, so make sure they describe the design concept clearly, step by step.
7. Ensure there are no gaps left in processes and procedures that could result in misinterpretation by others on the team.

Development

The development stage will differ according to the industry. ICT development requires a team of developers or coders who each work on one part of the system.

Engineering projects may include computer coding as well as physical construction using a variety of media that could include plastics, steel, wood, carbon fibre or other material appropriate to the task.

Business process and procedure development will normally involve writing, but can include ICT systems, Web pages and engineered devices, as well as posters, display boards, quick reference charts and product or procedure labelling. It may include multimedia, using video, film, computer graphics, sound and animation. It could also include painting arrows, feet or other signage on walkways and roads.

No matter what activities are included in the development stage, there will normally be a team of people taking part, each working on a discrete part of the project. These activities must be coordinated with a workflow process.

8. The specifications must always be followed to the letter.
9. Use a workflow process tool to coordinate the activities of the team.

Testing

Testing is vitally important and must be done thoroughly. There are two types of testing in common use, unit testing and whole system testing.

Unit testing describes testing a part of the project. The most common way of doing this is for developers, builders or writers to swap the part of the system they have been working on with a buddy and each test the work of the other. In an ICT system, it could be a procedure to accept and validate the login of a registered member of the group. An engineered project could involve the testing of a new kind of switch. A business project could involve the testing of a specific report.

Whole system testing, as the name suggests, means putting the individual parts together and testing the whole product from beginning to end. Normally, independent testers do this work.

Both types of testing require checking each part of the system against the specifications. If an anomaly is found, there must be a clear process for indicating the discrepancy, passing that part of the system back to the developers and checking again after retrofitting.

A serious problem with testing complex systems is the interactivity of the components. Because these systems can be seen as an organic whole, each component has an impact on other parts of the system. If a developer is asked to make a change to one component, testing of the whole system must begin again because that change may affect other parts of the whole.

A second major problem can also occur at the whole system testing stage. Testing, as has been said before, is carried out against the specifications. If there is a flaw in the specifications, that flaw will be mirrored in testing. For this reason, it is vitally important that end users participate in the design stages, particularly when forming the design concept and during whole system testing.

If end users have been involved in the formation of the design concept, there are less likely to be flaws carried through to the specifications writing stage. If they also participate in the whole system testing, they will not be testing against the specifications. They will be testing against their knowledge of operational requirements.

10. Unit testing can be done by either the development team or the testing team.
11. Whole system testing must be done by an independent team who test against the specifications.
12. Testing teams must include end users who can test against operational requirements.

Deployment

I have been present in workplaces where new systems, devices or procedures have been introduced with no warning and often with no training. As you can imagine, the new deployment is generally met with surprise and negativity. If the end users have had no part in the design, they will poke and prod to find fault. Commitment to the new product is in this case low.

On the other hand, I have worked in organisations where end users have participated in the design stages and have taken a leading role in implementation and training. The difference was like chalk and cheese. The new product was welcomed and acceptance and commitment high.

13. Involve end users in at least design concept decisions, testing and implementation of new products. Acceptance and commitment will be high from day 1.

Training

I have been amazed at the number of times I have witnessed the deployment of new systems and processes without any training. One manager answered my question by saying, 'They'll work it out on their own. They don't need training, besides it costs too much'.

The constant request from end users is for detailed training on new tools and processes. Training can take many forms. Occasionally, a training team will deliver face-to-face training. Sometimes, one or two staff members are trained, and they are expected to peer train their fellow workers. The other commonly used form of training involves developing training materials and posting them on a Website.

14. Whatever the form taken, training must be included in the implementation plan of new products and processes.

Maintenance

The last step in the design process, maintenance, is often ignored. This step is crucial where end users do not participate in the design process of their tools and procedures, but it is still very important where they do participate because situations change and constant updating is required in our fast-moving society.

I was called into a control centre as a consultant at a time when the controllers were having trouble with an upgraded ICT system that allowed them to remotely turn on and off high-voltage power lines and switches in sub-stations across the state. They had had no contact with the design team and no training on this upgrade, and were concerned that several vital operations seemed to do the opposite of what they expected. I was told that a programmer from the design company was on site but that he refused to talk with the controllers. I approached him and pointed out their concerns.

'It's not my job to train them or fix their problems. You'll have to talk to head office', was the reply.

I spoke to business managers in the organisation about securing a maintenance contract with the design company.

15. Ensure that a strategy for gathering end-user concerns is in place.
16. Make sure a maintenance arrangement is negotiated with third-party (or internal) design teams so that anomalies can be dealt with as quickly as possible.

DESIGN PROCESS AUDITING

Major causes of design error underlie the 16 rules outlined above, but as you can imagine, many more issues than these can have an impact on the potential of a system, device or process to contain latent hazards.

One group of issues relates to the constraints of design imposed on the designers. Constraints can be imposed by the design company or workgroup managers or by the client group, that is, those who have requested the design product. The common constraints are insufficient time, budget or resources, both human and material. The imposition of any of these constraints causes the design team to cut corners by reducing the project scope, lessening the detail of the project and often cutting short testing time, thereby reducing the thoroughness of the all-important testing phase before deployment of the design product.

How Many Ways Can the Design Process Go Wrong?

Whew! What a task we have in attempting to identify where we might find the elements that can lead to safety breeches, to errors that can cause accidents.

In our discussion about the disconnect model, we discovered that less than a quarter of all designers participating in my study discussed the new system with end users in a recent project. And we read that the very costly Queensland Health payroll system fiasco was berated in the second KPMG report (2012) for operating from a flawed design concept because no one had bothered to talk to anyone in the health department, let alone the payroll people, to discover the number of staff who had to be paid, and how many different salary scales had to be built into the system.

So discussions with end users during the design concept formation stage, in fact during all design stages, but particularly also during testing, are vitally important for a safe system. I have said elsewhere that the choice of an adaptive SDLC model to guide the design process is very wise, even mandatory for YELLOW CAP design projects. So given that we are convinced to include end users in discussions throughout the design process and be guided by an adaptive SDLC, we have reduced our error potential enormously. These two strategies will ensure a shorter development time and save money that would normally be spent on reworking. They will also reduce the potential design error risk substantially.

But, that is only the start of the story. Let us look at our designers again. Let us say one of the designers has a toothache. His head hurts and his concentration is affected. He is attempting to plan a complex part of the new system, but he misses a vital step. Here is a potential accident waiting to happen. He passes his work across to a fellow designer to check. She had a row with her boyfriend this morning and is still upset. She misses the vital step also. She passes the design on to the documenter who will write the specifications for this part of the project. He writes up the plan for this section of the project as a set of instructions for the builder. He does not understand the importance of the missed step. In fact, he does not know what has been missed, so his specifications do not include it. The builder builds what has been written up, and the tester tests against the specifications, so none of these people know about the missed step.

This kind of hazard happens every day and can be described clearly using the domino effect model or Reason's Swiss cheese model. Both of these are very good at explaining sequential linear events.

But, designers operate in an organisational climate. What if the boss tells the designers that the budget is limited for this particular project, a timer for a steam cleaning unit, and it needs to be built quickly? Also, there is no budget for a documenter, so the designer has to write the specifications. There is only a short opportunity for testing as well. The project is given to a young designer who has had no previous experience with timing mechanisms. Because of the hurry, he has little time to do any relevant research. He googles a design for a timer that seems it might do the job, and quickly writes up the specifications. The unit is built and given a cursory test before being deployed. The timer goes haywire, the steamer blows its top and the end user is severely scalded.

Here, we see not a linear progression of problems, but an interrelation of a number of factors, some of them organisational and some personal. This kind of scenario is more common than a linear progression of issues. For this, I have devised the random clusters model of accident causation that can explain the interrelationships between personal, technical, budgetary, temporal, resource-based and organisational factors.

So, we can examine each step in a design process and identify problems, but we have to remember that any system or process is an organic whole, made up of many small pieces, and a fault in one part of a design process may be magnified or negated by activities in another design process step.

ANALYTICAL TOOL

In order to help designers, managers and business and systems auditing staff to identify possible hazards in the design process, a new analytical tool is being constructed. The Design Error and Accident Mitigation (DEAM) tool will not only assess design processes and identify the interactions that might cause errors, but also provide a basis for training staff in error reduction strategies, and provide accident investigators with a tool to help them determine the causes of incidents and recommend strategies to avoid design error problems in the future.

DEAM is based on the results of a three-year study of the design processes used to construct new systems, devices and processes, and the interactions between designers, managers and end users as those new technologies, tools and work procedures were built and implemented. More information about this tool will be available in due course.

Glossary of Terms

adaptive SDLC: Set of principles guiding the development of a design project where the end point is not clearly known at the beginning of the process and stakeholders and end users form part of the design team and contribute every step of the way.

Agile: Perhaps the most popular of *adaptive* SDLCs.

artificial intelligence: Exhibiting of seemingly intelligent behaviour, normally seen only in humans, by technological systems, for example robots.

automated testing: Use of computer tools to test newly designed information and communication technology systems.

automation: When machines take over the operation of tools and processes replacing the humans who previously were the operators.

complexity levels: Levels of complexity in the design and development of a system, device or business process that increase the potential for design error.

continuous Agile: Modified SDLC where a continuous maintenance billing model is added to the *Agile* model to allow the system, device, or process design to remain open and ready to be modified at a moment's notice.

disconnect model: Model that explains the problems created for designers and the error potential when end users are not included in the formation of a design concept.

interlocking: Interrelational working of points, signals and other rail devices that ensure once a train route is set, it cannot be reversed until the designated train has passed by.

intuition: Kind of inner perception where the processes by which a 'knowing' occurs remain mostly unknown to the thinker.

IoT: Internet of Things: a new term to describe machine to machine interaction as sensors and systems become more complex and interrelated.

M2M: Machine to machine: a term that describes the way sensors and other sensing devices feed their readings to a computerised system, an example being the thermostat that reads the temperature and directs an air conditioner to raise or lower the temperature in an enclosed space.

Maglev railway track: Where a process called magnetic levitation is used to raise the wheels of a train a slight amount to lower friction and increase speed. Electromagnetic impulses produce a similar pole in track and wheels to create repulsion.

participative design: Where the end-user participates in the design process of tools or technologies he or she will use.

random clusters: New accident causation model that describes the way that a number of disassociated factors can combine to create new error potential in a complex system, device or process.

SDLC: Systems development life cycle model: a set of steps followed to plan and develop a solution to a problem. It originally started by guiding the steps in the design and development of an information technology system, but has been adopted by scientists, engineers and business design people as they develop new technologies and devices or business procedures and processes.

SME: Subject matter expert: a person who holds specialised knowledge in an area. That person will often act as an adviser to a design team supplying facts and insights they do not possess.

traditional SDLC: Set of processes guiding the development of a design such as the *waterfall* SDLC where the end point is clearly determined at the beginning of the project and there is minimal or no interaction between designers and end users during the design process.

waterfall: First systems development life cycle model where the design steps follow each other like water cascading down a waterfall.

whiteout: Condition that occurs in Antarctica when cloud layers diffuse the direct rays of the sun, making the terrain look like sea ice.

References

Australian Associated Press (AAP). 2011. Queensland Health payroll problems far from over. http://www.medicalsearch.com.au/qld-health-payroll-problems-far-from-over-says-minister/n/49774 (accessed March 4).

Abend, L. 2013. Spanish train crash mystery: Why didn't automated brakes stop the disaster? http://world.time.com/2013/07/26/ (accessed July 26).

Ambler, S. W. 2005. The agile system development life cycle (SDLC). Ambysoft. http://www.ambysoft.com/essays/agileLifecycle.html (accessed Feb 9).

Australian Bureau of Statistics. 2002. Level crossing accident fatalities by their mode of transport, 1997–2002. https://www.atsb.gov.au/media/33881/level_cross_fatal.pdf (accessed Nov 4).

Beck, K. and C. Andres. 2004. *Extreme Programming Explained: Embrace Change*. 2nd ed., XP Series. Boston: Addison-Wesley.

Besnard, D. and G. Baxter. 2003. Human compensations for undependable systems. Technical Report Series. Newcastle upon Tyne, UK: University of Newcastle upon Tyne.

Boehm, B. 1986. A spiral model of software development and enhancement. *Computer* 21 (5):61–72.

Brown, M. 2008. *Erebus: The Air New Zealand Crash, 1979, Australia's Worst Disasters*. Sydney: Hachette Australia.

Casey, S. 1998. *Set Phasers on Stun – And Other True Tales of Design, Technology and Human Error*. 2nd ed. Santa Barbara, CA: Agean.

Chapanis, A. 1996. *Human Factors in Systems Engineering*. Wiley Series in Systems Engineering. Toronto: Wiley.

Currie, J. R. L. 1971. *The Runaway Train: Armagh 1889*. Newton Abbot, UK: David & Charles.

Dekker, S. 2006. *The Field Guide to Understanding Human Error*. Surrey, UK: Ashgate.

Dekker, S. 2011. *Drift into Failure: From Hunting Broken Components to Understanding Complex Systems*. Surrey, UK: Ashgate.

Foster, C. 2014. How M2M is changing the mobile economy. Gigaom Research. https://gigaom.com/report/how-m2m-is-changing-the-mobile-economy/ (accessed May 14).

Halcrow. 2010. *Independent Review: Failures of QR Universal Traffic Control*. Brisbane: Halcrow Pacific.

Hangar, I. 2014. Royal Commission into the Home Insulation Program. Australia. http://www.homeinsulationroyalcommission.gov.au/Pages/default.html (accessed May 15).

Harding, L. 2006. At least 23 die as driverless train crashes into maintenance truck in Berlin. *The Guardian*. http://www.theguardian.com/world/2006/sep/23/germany.topstories3 (accessed September 23).

Heinrich, H. W. 1931. Industrial accident prevention: A scientific approach in. In Hollnagel, E., *2009 Safer Complex Industrial Environments: A Human Factors Approach*, p. 10. Boca Raton, FL: CRC Press.

Hollnagel, E. 2006. Resilience: The challenge of the unstable. In Hollnagel, E., Woods, D. D. and Leveson, N. (eds.) *Resilience Engineering*. UK: Ashgate.

The Hindu. 2011. Report on the high speed rail crash near Wenzhou, China. Delhi, July 28.

IBM. 2012. Applying agile principles to the development of smarter products. Thought Leadership White Paper, October. Armonk, NY: IBM.

Johansen, T. and T. Gilb. 2005. *From Waterfall to Evolutionary Development (Evo): How We Rapidly Created Faster, More User-Friendly and More Productive Software Products for a Competitive Multi-National Market.* Somerset, UK: INCOSE.

Kay, R. 2002. QuickStudy: System development life cycle. *Computerworld*, May 14.

KPMG. 2012. Review of the Queensland Health payroll system. Brisbane: Queensland Health.

Mansfield, I. 2012. A trip in Heathrow's driverless transit system. Transport Issues. https://www.ianvisits.co.uk/blog/2012/08/26/a-trip-in-heathrows-driverless-transit-system/ (accessed August 26).

May, E. L. and B. A. Zimmer. 1996. The evolutionary development model for software. *Hewlett-Packard Journal* 1 (8).

McDonald, W. 2000. Restoring the Nullarbor: The application of vital telemetry to self-restoring points across the Nullarbor. *IRSE*, July.

Norman, D. A. 1988. *The Design of Everyday Things.* New York: Basic Books.

Norman, D. A. 2007. *The Design of Future Things.* New York: Basic Books.

Norman, D. A. 2011. *Living with Complexity.* Cambridge, MA: MIT Press.

Persaud, D. 2015. Machine-to-machine technology automates the warehouse. TechTarget. http://searchmanufacturingerp.techtarget.com/feature/Machine-to-machine-technology-automates-the-warehouse (accessed July 15).

Qureshi, Z. H. 2007. A review of accident modelling approaches for complex socio-technical systems. In *12th Australian Workshop on Safety Related Programmable Systems (SCS'07)*, Adelaide.

Reason, J. 1995. A system approach to organizational error. *Ergonomics* 38:1708–1721.

Reason, J. 2008. *The Human Contribution.* Surrey, UK: Ashgate.

Ruvidini, A. 1999. Collision Indian Pacific passenger train 3AP88 and freight train 3PW4N, Zanthus, WA. Independent Investigation Report. Perth: Department of Transport, WA Government.

Stephens, M. 2008. The case against extreme programming - A self-referential safety net. http://www.softwarereality.com/lifecycle/xp/index.jsp (accessed November 12).

Tevell, E. and M. Ahsberg. 2011. Positive and negative quality effects in distributed scrum projects – An industrial case study. Master of science, University of Gothenburg, Sweden.

Vicente, K. J. 1999. *Cognitive Work Analysis.* London: Lawrence Erlbaum Associates.

Yarrow, R. 2014. Driverless cars already on the road. *Telegraph.* http://www.telegraph.co.uk/motoring/car-manufacturers/volvo/10804595/Driverless-cars-already-on-the-road.html (accessed May 6).

Index